# Solidification and Casting of Metals and Alloys

# Solidification and Casting of Metals and Alloys

Guest Editor

**Wenchao Yang**

MDPI

Basel • Beijing • Wuhan • Barcelona • Belgrade • Novi Sad • Cluj • Manchester

*Guest Editor*
Wenchao Yang
State Key Laboratory of
Solidification Processing
Northwestern Polytechnical
University
Xi'an
China

*Editorial Office*
MDPI AG
Grosspeteranlage 5
4052 Basel, Switzerland

This is a reprint of the Special Issue, published open access by the journal *Metals* (ISSN 2075-4701), freely accessible at: https://www.mdpi.com/journal/metals/special_issues/91GYRCOOZK.

For citation purposes, cite each article independently as indicated on the article page online and as indicated below:

| Lastname, A.A.; Lastname, B.B. Article Title. *Journal Name* **Year**, *Volume Number*, Page Range. |
|---|

**ISBN 978-3-7258-6446-1 (Hbk)**
**ISBN 978-3-7258-6447-8 (PDF)**
**https://doi.org/10.3390/books978-3-7258-6447-8**

# Contents

# About the Editor

**Wenchao Yang**

Wenchao Yang is a Tenured Professor at the School of Materials Science and Engineering/State Key Laboratory of Solidification Processing, Northwestern Polytechnical University in China. His research interests lie in the development, solidification, and mechanical properties of nickel-based superalloys and lightweight metallic materials such as aluminum alloys and aluminum matrix composites. His research integrates artificial intelligence, computer modeling in automated design, experimental investigations of structure and properties at laboratory scale, and manufacturing applications. He has published over 110 SCI-indexed papers in international and leading domestic journals in metallic materials science, such as the *International Journal of Plasticity*, *Journal of Materials Science & Technology*, *Materials Research Letters*, and *Scripta Materialia*. In addition to this, he currently holds 22 authorized national invention patents. He is a recipient of the National Science Fund for Excellent Young Scholars and was admitted to the Young Talent Support Program of the China Association for Science and Technology. He has also received an award from the Shaanxi Provincial Outstanding Young Science Fund and is recognized as one of the "100 Scientific and Technological Innovation Talents" in Hunan Province in China. He is the recipient of several other awards, such as the China Industry-University-Research Cooperation Innovation Award, the Young Scientist Award of the Chinese Materials Research Society, the Top Academic Paper Award of China's Excellent Sci-Tech Journals, the First Prize of Shaanxi Provincial Science and Technology Award, and the Young Science and Technology Award of the Shaanxi Society of Aeronautics.

*metals*

MDPI

*Editorial*

# Solidification and Casting of Metals and Alloys

**Wenchao Yang**

State Key Laboratory of Solidification Processing, Northwestern Polytechnical University, Xi'an 710072, China; wenchaoyang@nwpu.edu.cn

## 1. Introduction and Scope

Establishing control of the cast structure is the underlying object of solidification metallurgy [1,2]. Recent advances and developments in solidification are enabling the production of high-purity castings for superalloys, aluminum, magnesium, titanium and copper alloys, rapidly solidified structural components and castings with unique microstructures [3,4]. Additionally, recent developments in processing technology allow for better productivity and reliability in some metal castings [5–8]. These developments have all stemmed from a good understanding of the science of solidification metallurgy, as well as an appreciation of the merits of structural control using advanced solidification processing [9–12]. In addition, numerous works have been developed via both experimental and analytical/computer modeling for disclosing the fundamental aspects of the metallurgical process, phase formation and growth within liquid melts, and many calculated software have been used to clarify solidification. These include ProCAST, AnyCasting, Magma, Micress, etc. [13–16].

Accordingly, this Special Issue of *Metals*, entitled "Solidification and Casting of Metals and Alloys", will review the latest developments in the various aspects of solidification metallurgy. Specifically, we will cover the following: (a) metallurgical control of the composition and microstructure of metals or castings; (b) micro- and macrosegregation mechanisms, as well as the microstructural evolution of solidification microstructures; (c) multi-scale experiments and simulations for solidification using different calculated methods; (d) fundamental aspects such as nucleation, grain growth and the development of the mushy zone; (e) thermal, compositional effects on the development/avoidance of casting defects, etc.

## 2. Overview of the Published Articles

The present Special Issue includes ten research studies: one review and nine articles. These studies cover superalloys, iron/steel, Al alloys and so on, focusing on microstructure–property relationships in the field of metals and alloys.

Lihui Zhang et al. (Contribution 1) reported on the formation mechanism of oxide inclusions in a high-aluminum Ni-based superalloy during the vacuum induction remelting process. In their study, the oxide inclusions in Ni-based superalloys play a crucial role in determining their mechanical properties, oxidation resistance and corrosion resistance at high temperatures. In the study, the sources and formation mechanisms of different types of oxide inclusions in a high-aluminum Ni-based superalloy were systematically studied. An automatic field emission scanning electron microscope equipped with an energy-dispersive spectrometer and a self-designed superalloy inclusion analysis standard was utilized to quantitatively reveal the oxide inclusion characteristics of the high-aluminum Ni-based superalloy prepared via vacuum induction melting (VIM) and vacuum induction remelting (VIR) processes. The experimental results indicate that the typical oxide inclusions in



https://doi.org/10.3390/met15010087

the Ni-based superalloy before the VIR process are irregular $MgO{\cdot}Al_2O_3$ inclusions with sizes of less than 2 μm. After the VIR process, the typical oxide inclusions in the Ni-based superalloy are also $MgO{\cdot}Al_2O_3$ inclusions. However, these oxide inclusions can be classified into three categories: (i) endogenous irregular $MgO{\cdot}Al_2O_3$ inclusions, less than 4.3 μm in size, inherited from the master alloy; (ii) several hundred-micron film-like $MgO{\cdot}Al_2O_3$ inclusions generated as interface reaction products between the MgO crucible and melts; (iii) millimeter-scale $MgO{\cdot}Al_2O_3$ inclusions and several tens of microns of MgO inclusions from the exfoliation of the MgO crucible matrix.

Richard Sheridan et al. (Contribution 2) reported on the strip casting of $Sm_2TM_{17}$-type alloys for the production of the metastable $SmTM_7$ Phase. In their study, the conventional book casting of $Sm_2TM_{17}$-type alloys (where TM = Co, Fe, Cu, Zr) leads to a coarse, highly segregated microstructure, predominantly due to the slow, variable cooling rate from the mold surface towards the center of the ingot. These cast alloys require a long homogenization treatment to remove this segregation and develop a super-saturated, metastable $SmTM_7$-type hexagonal phase. This $SmTM_7$ phase is a vital precursor phase required during magnet production to develop the complex cellular structure responsible for high magnetic properties. In the study, strip casting was employed to facilitate rapid solidification to develop thin flakes (<0.5 mm thick) with a columnar grain structure. Rapid cooling has the potential to produce a homogenous microstructure consisting predominantly of the metastable $SmTM_7$ phase. This could remove or significantly reduce the need for the energy-intensive homogenization treatment usually required in conventional magnet manufacture. This study investigated the effect of wheel speed (and hence cooling rate) on the flake thickness, microstructure and phase balance of the cast alloys. It was shown that for wheel speeds between 1.1 and 3.0 m/s, the microstructure showed large variation; however, in all cases, evidence of the columnar $SmTM_7$ phase was presented. The adhesion between the melt and the wheel was deemed to be critical for the nucleation and subsequent columnar growth of $SmTM_7$ grains, where the wheel speed controlled both the flow of the alloy onto the wheel and the thickness of the resultant flake. It was determined that in order to achieve a homogenous columnar $SmTM_7$ structure, the maximum flake thickness should be limited to 270 μm to avoid the formation of equiaxed $Sm_2TM_{17}$ grains through insufficient cooling.

María Carmen Manjabacas et al. (Contribution 3) reported a methodology used to define the Niyama criterion reinforced with solid fraction analysis: this method was applied to the sand casting of steel bars. In their study, Niyama and solid fraction criteria were used to predict the solidification porosity and microporosity in the computing simulation of casting processes. The solid fraction permits us to determine the areas that solidify last and that are candidates for presenting porosity if a feeding system is not correctly designed. The Niyama criterion is locally obtained based on the thermal and cooling gradients at a point of the liquid casting. The Niyama value at a casting point varies rapidly from low rates to high ones during the last part of metal solidification, which demands that the percentage solidification of the metal is defined to determine the Niyama number. In addition, the Niyama threshold that establishes the soundness of the workpiece can vary according to the nature of the metal or the casting system. In the study, a methodology to determine the solidification percentage was presented. The method was based on Niyama number evolution during the solidification process at different key points. These points were validated by the solid fraction criterion as healthy or, on the contrary, as candidates for containing porosity. In addition, some considerations of the solid fraction criterion were visited since the threshold value for which the isolation of the last solidification areas can be defined was not clear. The study was validated by the empirical casting criteria presented in the literature for obtaining sound parts and applied to low-carbon steel bars produced via sand casting.

Hongliang Ma et al. (Contribution 4) reported the simulation of the mold filling and solidification characteristics of hypereutectic high-chromium cast iron via lost foam suspension casting. In their study, the finite element software ProCAST version 14.5 was used to simulate the temperature field, flow field and defect prediction in the filling and solidification process of hypereutectic high-chromium cast iron. The effects of pouring temperature, negative pressure and the amount of suspension agent added during the technological process were explored. The optimum process parameters were presented. It was found that the suspension agent had a certain hindrance to the filling process, but the filling process remained stable. In the solidification stage, 89.4% of the suspension agent melted, resulting in a relative supercooling degree of 50 °C, which had a certain chilling effect and improved the solidification rate.

Zhikun Ma et al. (Contribution 5) reported the microstructure and anisotropy of mechanical properties of Al-3Li-1Cu-0.4Mg-0.1Er-0.1Zr alloys prepared via normal rolling and cross-rolling. In their study, the influence of normal rolling and cross-rolling on the microstructure, mechanical properties and anisotropy of Al-3Li-1Cu-0.4Mg-0.1Er-0.1Zr alloy was investigated. With an increase in the rolling reduction amount, both the strength and plasticity of the alloy were enhanced. Among them, the alloy in the normal rolling state with a deformation amount of 90% exhibited the best properties, with a tensile strength of 362 MPa and an elongation of 19.1% along the rolling direction. During the rolling process, the intergranular Cu-containing phase in the alloy was continuously broken and dissolved, decreasing in both size and quantity, shifting from continuous distribution along grain boundaries to a granular distribution. Moreover, a large quantity of the Al3Li phase and $Al_3$(Er, Zr, Li) core–shell composite phase were precipitated in the alloy. Recrystallization occurred mainly through the particle stimulated nucleation (PSN) mechanism. Cross-rolling eliminated the brass-type texture <111> produced via normal rolling and enhanced the brass R-type texture {111}<112>. The index of plane anisotropy (IPA) of the strength decreased from 10.1% for normal rolling to 5.5% for cross-rolling, and the IPA of elongation decreased from 12.8% to 3.3%. Cross-rolling provided an effective method to reduce the anisotropy of Al-Li alloys.

Miao Huo et al. (Contribution 6) reported on the stray grains from fragments in the rejoined platforms of Ni-Based single-crystal superalloy. In their study, in order to study the origin of misorientation defects during solidification, a model with rejoined platforms was designed according to the geometry of single-crystal guide vanes. Electron Back-Scattering Diffraction (EBSD) was used to quantify the orientation deviation of the dendrites and identify the solidification defects in the rejoined platforms. The results showed that stray grain defects appeared in the platforms and that their misorientation changed gradually, not abruptly. Combined with the simulation results, it was proposed that the stray grains formed as the result of dendrite fragmentation, which was induced via solute enrichment in the mushy zone during solidification. Meanwhile, it was accompanied by an obvious dendritic deformation, which was caused by solidification shrinkage stress. This suggested that the fragmentation was induced by multiple factors, among which the concave interface shape provided favorable conditions for solute enrichment, and the dynamic variability in the local thermal gradient and fluctuations of the solidification rate might have played catalytic roles.

Jordan Roger Kozakevich et al. (Contribution 7) reported the solidification kinetics of an Al-Ce Alloy with additions of Ni and Mn. Their study focused on characterizing the solidification of an Al-Ce alloy with additions of Ni and Mn (nominal composition Al-12.37Ce-3.26Ni-0.94Mn-0.12Fe in weight percentage). The alloy was cast in a wedge mold configuration, resulting in cooling rates between 0.18 and 14.27 °C/s. Scanning electron microscopy coupled with the energy-dispersive X-ray spectroscopy and differential scanning calorimetry techniques characterized the evolution rates of solid phases. The SEM/EDS data revealed that an $Al_{10}CeMn_2$ phase was present at higher cooling rates. At lower cooling rates, near the center of the casting, a primary $Al_{23}Ce_4Ni_6$ phase was more present. It was observed that up to 2.6 atomic percent (at.%) of Mn was dissolved in this primary $Al_{23}Ce_4Ni_6$ phase, thereby removing a large portion of the available Mn for forming the Al10CeMn2 phase. DSC analysis showed differences in the samples' liquidus temperatures, which indicated compositional variations. Inductively coupled plasma–atomic emission spectroscopy and Scheil solidification simulations correlated the compositional differences with phase formation, which agreed with the SEM and DSC results. This experiment provided insight into novel Al-Ce-Ni-Mn alloys and where their potential lies in industrial applications.

Zhixian Peng et al. (Contribution 8) reported on the cracking behavior and high-temperature thermoplastic analysis of 09CrCuSb steel billets. This study characterized the surface crack morphology and elements of 09CrCuSb steel continuous casting slabs and studied the high-temperature thermoplasticity of continuous casting slabs using a Gleeble 3500 thermal simulator. Combining the macro/micro-fracture morphology and microstructure characteristics, the formation rules of slab cracks were discussed. The research results show that the increase in Ae3 temperature caused by changes in alloy elements resulted in the precipitation of a thin, film-like coexisting ferrite along the grain boundaries at a certain temperature, as well as the element segregation behavior of low-melting-point alloy elements at the original austenite grain boundaries, which were the main factors inducing cracks in 09CrCuSb steel continuous casting billets. The plasticity of 09CrCuSb steel at high temperatures was poor at 800~900 °C. In continuous casting process control, it is necessary to try to avoid keeping the billet in this temperature range long term.

Zhiguo Li et al. (Contribution 9) reported the grain refinement of aluminum and aluminum alloys by Sc and Zr. Rare earth elements and transition elements are widely used in aluminum alloys, magnesium alloys and copper alloys due to their unique microalloying effects. With regard to in-depth research on the grain refinement characteristics of rare earth elements and transition elements, the combination of grain refinement and microalloying in master alloys has a great impact on the theories and technical research of refinement and the broadening of the application field of master alloys. The study first summarized the grain refinement mechanisms and analyzed the effects of rare earth elements and transition elements on the grain refinement of aluminum and aluminum alloys, and it then summarized the elements suitable for the preparation of Al-M master alloys and their refinement mechanism.

Mohammed Y. Abdellahi et al. (Contribution 10) reported on the experimental evaluation of the mechanical and tribological properties of segregated Al-Mg-Si alloy filled with alumina and silicon carbide through different types of casting molds. In their study, two components of the 6061aluminum alloy were characterized by high mechanical and abrasive strength. The aims of the study were to understand the effects of different types of cooling rates through different mold materials and to investigate the effect of casting with ceramic additives on segregation of the aluminum alloy itself as a composite material forum. Therefore, a series of mechanical tests were conducted, such as a compression test, Vickers hardness and pin-on-disk wear test. The samples were cast at 650 °C and in electric furnaces for 2 h to ensure that the metal achieved adequate homogeneity and temperature. Then, abrasive macroparticles of $Al_2O_3$ and SiC with a size close to 40–60 μm were used. The particles were poured under constant stirring for 1 min. Then, they were cast in two types of molds: steel and graphite. The cast specimens were obtained as a reference without particles and with 0.5 wt.%, 1 wt.%, 2 wt.%, 3 wt.%, 4 wt.% and 8 wt.%. The thermal effect and the heat due to conduction and radiation were calculated. The maximum compressive strength increased by ≈21% with SiCp cast in graphite molds, and HV increased by ≈29% with SiC cast in graphite molds. The same was found for wear resistance, which became good with SiC cast in graphite molds, and it was generally found that the cooling rate through the mold weakened the alloy due to the segregation effect. The presence of tough particulate through the aluminum matrix barrier created a number of loads. Additionally, the high specific heat of graphite, which plays a dominant role in the slaw cooling rate of casting, led to grain enlargement, whereas the higher cooling rate of steel led to grain refinement. These concepts are the main rules of heat treatments through the casting process, and they save time and effort.

**Acknowledgments:** We are grateful to all staff at MDPI for their valuable collaboration. We would like to express our gratitude to all of the contributing authors and reviewers; without their excellent work, it would not have been possible to publish this Special Issue, which we hope will be both interesting and of lasting importance in general reading and the reference literature.

**Conflicts of Interest:** The author declares no conflicts of interest.

**List of Contributions:**

1. Zhang, L.; Liu, E.; Xing, W.; Xue, Z.; Fan, W.; Zhao, Y.; Luo, Y.; Ge, C.; Xia, M. The Formation Mechanism of Oxide Inclusions in a High-Aluminum Ni-Based Superalloy during the Vacuum Induction Remelting Process. *Metals* **2024**, *14*, 654. https://doi.org/10.3390/met14060654.
2. Sheridan, R.; Gresle-Farthing, J.; Appleby, A.; Brown, M. Strip Casting of $Sm_2TM_{17}$-Type Alloys for Production of the Metastable SmTM7 Phase. *Metals* **2024**, *14*, 517. https://doi.org/10.3390/met14050517.
3. Manjabacas, M.C.; Miguel, V. A Methodology to Define the Niyama Criterion Reinforced with the Solid Fraction Analysis: Application to Sand Casting of Steel Bars. *Metals* **2023**, *13*, 1777. https://doi.org/10.3390/met13101777.
4. Ma, H.; Fu, H. Study on Simulation of Mold Filling and Solidification Characteristics of Hypereutectic High-Chromium Cast Iron by Lost Foam Suspension Casting. *Metals* **2023**, *13*, 1761. https://doi.org/10.3390/met13101761.
5. Ma, Z.; Zhong, T.; Sun, D.; Qian, B.; Turakhodjaev, N.; Betsofen, S.; Wu, R. Microstructure and Anisotropy of Mechanical Properties of Al-3Li-1Cu-0.4Mg-0.1Er-0.1Zr Alloys Prepared by Normal Rolling and Cross-Rolling. *Metals* **2023**, *13*, 1564. https://doi.org/10.3390/met13091564.
6. Huo, M.; Chen, C.; Jian, H.; Yang, W.; Liu, L. The Stray Grains from Fragments in the Rejoined Platforms of Ni-Based Single-Crystal Superalloy. *Metals* **2023**, *13*, 1470. https://doi.org/10.3390/met13081470.
7. Kozakevich, J.R.; Stroh, J.; Sediako, D.; Weiss, D. Solidification Kinetics of an Al-Ce Alloy with Additions of Ni and Mn. *Metals* **2023**, *13*, 955. https://doi.org/10.3390/met13050955.
8. Peng, Z.; Mei, T.; Zheng, J.; Yuan, Y.; Wang, L. Cracking Behavior and High-Temperature Thermoplastic Analysis of 09CrCuSb Steel Billets. *Metals* **2023**, *13*, 1058. https://doi.org/10.3390/met13061058.
9. Lei, Z.; Wen, S.; Huang, H.; Wei, W.; Nie, Z. Grain Refinement of Aluminum and Aluminum Alloys by Sc and Zr. *Metals* **2023**, *13*, 751. https://doi.org/10.3390/met13040751.
10. Abdellah, M.Y.; Fadhl, B.M.; El-Ainin, H.M.A.; Hassan, M.K.; Backar, A.H.; Mohamed, A.F. Experimental Evaluation of Mechanical and Tribological Properties of Segregated Al-Mg-Si Alloy Filled with Alumina and Silicon Carbide through Different Types of Casting Molds. *Metals* **2023**, *13*, 316. https://doi.org/10.3390/met13020316.

## References

1. Dantzig, J.A.; Rappaz, M. *Solidification*; EPFL Press: Lausanne, Switzerland, 2009.
2. Verma, S.; Dewan, A. Solidification Modeling: Evolution, Benchmarks, Trends in Handling Turbulence, and Future Directions. *Metall. Mater. Trans. B* **2014**, *45*, 1456–1471. [CrossRef]

3. Niu, W.; Huang, S.; Lin, B.; Li, J. Study of the Solidification Microstructure and Deformation Behaviour of $Cu_{20}Fe$ Alloy. *Metals* **2024**, *14*, 1313. [CrossRef]
4. Ricci, S.; Iannitti, G. Mechanical Behavior of Additive Manufacturing (AM) and Wrought $Ti_6Al_4V$ with a Martensitic Microstructure. *Metals* **2024**, *14*, 1028. [CrossRef]
5. Li, Y.; Wang, L.; Wang, Z.; Li, J.; Wang, J. A phase-field model bridging near-equilibrium and far-from-equilibrium alloy solidification. *Acta Mater.* **2025**, *284*, 120596. [CrossRef]
6. Kadoi, K.; Matsumoto, Y.; Chiba, H.; Inoue, H. Solidification cracking susceptibility of alloy 718 during additive manufacturing and evaluating method. *J. Mater. Res. Technol.* **2024**, *33*, 6389–6396. [CrossRef]
7. Ye, Z.; Zhao, K.; Yu, Z.; Prashanth, K.G.; Zhang, F.; He, Y.; Peng, Y.; Wu, W.; Tan, H. Understanding the solute segregation and redistribution behavior in rapidly solidified binary Ti-X alloys fabricated through non-equilibrium laser processing. *Addit. Manuf.* **2024**, *96*, 104561. [CrossRef]
8. Yu, W.; Bu, Y.; Yao, J.; Zhang, S.; Zhao, Y.; Huang, X.; Ni, W. Leaching behavior and stabilization/solidification characterization of heavy metals from silicomanganese slag. *J. Hazard. Mater.* **2025**, *485*, 136915. [CrossRef]
9. Chandra, S.; Radhakrishnan, J.; Huang, S.; Wei, S. Ramamurty. Solidification in metal additive manufacturing: Challenges, solutions, and opportunities. *Prog. Mater. Sci.* **2025**, *148*, 101361. [CrossRef]
10. Li, H.; Wang, X.; Wang, J.Y.; Zhai, W.; Wei, B. Strengthening Cu-containing high entropy alloys through ultrasonic solidification constructed heterostructures. *Mater. Sci. Eng. A* **2024**, *916*, 147336. [CrossRef]
11. Han, J.; Luo, S.; Liu, N.; Chen, K.; Xie, S.; Wang, W.; Zhu, M. Experimental investigation on effect of cooling rate on carbide precipitation during solidification of high manganese steel. *J. Mater. Res. Technol.* **2024**, *33*, 1075–1086. [CrossRef]
12. Chen, L.; Wang, J.; Wang, Q.; Zhao, S.; Yu, T.; Zhao, Y. Multi-scale and multi-field coupling simulations of rapid melting and solidification process during laser DED Ni-based high-temperature alloy. *Int. J. Heat Mass Transf.* **2025**, *239*, 126538. [CrossRef]
13. Gong, T.; Hao, W.; Fan, W.; Chen, Y.; Chen, X.; Li, D. Accelerating phase-field simulation of multi-component alloy solidification by shallow artificial neural network. *Comput. Mater. Sci.* **2025**, *248*, 113594. [CrossRef]
14. Bodaghi, F.; Movahedi, M.; Na, S.J.; Zhang, L.J.; Kokabi, A.H. Effect of welding current and speed on solidification cracking susceptibility in gas tungsten arc fillet welding of dissimilar aluminum alloys: Coupling a weld simulation and a cracking criterion. *J. Mater. Res. Technol.* **2024**, *30*, 4777–4785. [CrossRef]
15. Seydani, M.Z.; Krimi, A.; Khelladi, S.; Bedel, M.; Mansor, M.E. 3D numerical simulation and experimental validation of resin-bonded sand gravity casting: Filling, cooling, and solidification with SPH and ProCAST approaches. *Therm. Sci. Eng. Prog.* **2024**, *47*, 102329. [CrossRef]
16. Saadati, A.; Malekan, M. Static solid cooling (SSC) method for directional solidification and single crystal production: A CAFE simulation approach. *J. Mater. Res. Technol.* **2024**, *33*, 9898–9910. [CrossRef]

metals

MDPI

Article

# The Formation Mechanism of Oxide Inclusions in a High-Aluminum Ni-Based Superalloy during the Vacuum Induction Remelting Process

Lihui Zhang [1,†], Erkang Liu [2,†], Weijie Xing [1], Zhaojiang Xue [2], Wenjie Fan [2], Yunsong Zhao [1,*], Yushi Luo [1], Changchun Ge [2] and Min Xia [2,*]

1 Science and Technology on Advanced High Temperature Structural Materials Laboratory, Beijing Institute of Aeronautical Materials, Beijing 100095, China; lihui.zhang@biam.ac.cn (L.Z.); xingwj621@163.com (W.X.); luoyushi1978@sohu.com (Y.L.)

2 Institute of Powder Metallurgy and Advanced Ceramics, University of Science and Technology Beijing, Beijing 100083, China; liu18203224277@163.com (E.L.); zhaojiang_xue@163.com (Z.X.); wejiefan2022@163.com (W.F.); ccge@mater.ustb.edu.cn (C.G.)

* Correspondence: yunsong.zhao@biam.ac.cn (Y.Z.); xmdsg@ustb.edu.cn (M.X.)

† These authors contributed equally to this work.

**Abstract:** Oxide inclusions in Ni-based superalloys play a crucial role in determining their mechanical properties, oxidation resistance, and corrosion resistance at high temperatures. In this paper, the source and formation mechanism of different types of oxide inclusions in a high-aluminum Ni-based superalloy were systematically studied. An automatic field emission scanning electron microscope equipped with an energy dispersive spectrometer and a self-designed superalloy inclusion analysis standard was utilized to quantitatively reveal the oxide inclusion characteristics of the high-aluminum Ni-based superalloy prepared via vacuum induction melting (VIM) and vacuum induction remelting (VIR) processes. The experimental results indicate that the typical oxide inclusions in the Ni-based superalloy before the VIR process are irregular $MgO \cdot Al_2O_3$ inclusions with sizes of less than 2 μm. After the VIR process, the typical oxide inclusions in the Ni-based superalloy are also $MgO \cdot Al_2O_3$ inclusions. However, these oxide inclusions can be classified into three categories: (i) endogenous irregular $MgO \cdot Al_2O_3$ inclusions, less than 4.3 μm in size, inherited from the master alloy; (ii) several hundred-micron film-like $MgO \cdot Al_2O_3$ inclusions generated as interface reaction products between the MgO crucible and melts; and (iii) millimeter-scale $MgO \cdot Al_2O_3$ inclusions and several tens of microns of MgO inclusions from the exfoliation of the MgO crucible matrix.

**Keywords:** oxide inclusion; formation mechanism; interface reaction; vacuum induction remelting; Ni-based superalloy

## 1. Introduction

Ni-based superalloys have been extensively applied in the aerospace industry due to their excellent mechanical properties, oxidation resistance, and corrosion resistance at high temperatures [1–3]. Investment casting is widely used in the manufacturing of high-temperature alloy components. During the alloy melting process, the contact between the high-temperature alloy liquid and refractory materials inevitably produces inclusions. Inevitably present in superalloy components, non-metallic inclusions serve as a key indicator for measuring the performance of Ni-based superalloys [4]. These seemingly small non-metallic inclusions are likely to be the nucleation locations of fatigue cracks, leading to the failure of superalloy components and even threatening people's lives [5–7]. Therefore, the characteristics of non-metallic inclusions in superalloys, such as their morphology, number, and distribution, should be seriously considered.

The common non-metallic inclusions in Ni-based superalloys are oxides, sulfides, carbides, etc., whereas oxide inclusions have received extensive attention [8–10]. In Ni-based superalloys, oxide inclusions may seriously affect the safety performance of superalloy components due to the large differences in the elastic modulus, deformation ability, and high temperature–oxidation resistance between the oxide inclusion and the matrix [11,12]. In October 2016, a high-pressure turbine disk made of an Inconel 718 alloy on the US B767-300 aircraft broke during takeoff at Chicago O'Hare International Airport. The investigation indicated that a large-sized film-like $MgO{\cdot}Al_2O_3$–TiN inclusion cluster in the turbine disk became the initiation location of a fatigue crack, which ultimately caused the accident [13]. Nevertheless, not every oxide inclusion in nickel-based superalloys necessarily causes the fracture of superalloy components. To some extent, small inclusions can promote the refinement of the microstructure and improve the strength of high-temperature alloys during solidification. Only larger inclusions can induce the initiation of cracks and even fractures in the superalloy matrix. A critical oxide inclusion size exists, below which there is no reduction in fatigue life resulting from crack initiation at oxide inclusions, and above which there is a transition from slip-induced to inclusion-induced crack nucleation [14]. In summary, it is crucial to control the size of oxide inclusions in nickel-based superalloys.

Vacuum induction remelting is a method of secondary remelting to produce high-quality superalloy components [15]. As the final step before casting, the VIR process directly affects the characteristics of oxide inclusions in superalloy components. There are many ways to introduce oxide inclusions into Ni-based superalloys during the VIR process, for example, (i) in the quality of the master alloys for the remelting of Ni-based superalloys. There is no doubt that a low-purity master alloy is bound to introduce a large number of oxide inclusions into the Ni-based superalloy alloy melt [16]. (ii) In interface interactions between the crucible and superalloy melt, the interface interactions include the physical dissolution of the crucible inner wall by the melt and the chemical dissolution of the crucible particle in the melt [17]. Previous research has shown that high-melting-point inclusions are introduced when using an $Al_2O_3$ crucible to melt Ni-based superalloys [18,19]. (iii) Within the remelting parameters of Ni-based superalloys, the evidence presented thus far supports the idea that the vacuum level [20], smelting time [21], casting temperature [22], etc., can significantly affect the composition of the superalloy melt. In turn, this indirectly affects the oxide inclusions in the superalloy's components.

In this work, systematic research was meticulously conducted to explore the origins of oxide inclusions in a typical high-aluminum nickel-based superalloy during the VIR process. By employing thermal simulation smelting experiments combined with comprehensive thermodynamic calculations, this study investigated the formation mechanisms of oxide inclusions within a high-aluminum Ni-based superalloy. The insights gained from these calculations are aimed at elucidating the complex interactions and conditions that lead to the presence of oxide inclusions. It is anticipated that the findings from this research will significantly contribute to the broader scientific understanding of the potential formation mechanisms and offer strategic approaches for controlling oxide inclusions in Ni-based superalloys.

## 2. Materials and Methods

### 2.1. Experimental Materials

The experimental material in this study is a master high-aluminum Ni-based superalloy produced by the VIM process. In the VIR process, a 20 kg master alloy undergoes melting within a vacuum induction furnace set at 1873 K, with a vacuum level maintained at 0.5 Pa and a melting time of 20 min. Subsequently, the molten alloy is gradually cooled to 1823 K and held at this temperature for 3 min to ensure homogeneity and alleviate any remaining impurities. Following this refining stage, the refined metal is then carefully poured into a prepared mold, where it solidifies into its final form. Table 1 lists the nominal composition of the Ni-based superalloy before and after the VIR process. It is worth mentioning that the content of the active element Al in the Ni-based superalloy is 2–5%.

**Table 1.** The nominal composition of the high-aluminum Ni-based superalloy (wt. %).

| Sample | Cr | Co | W | Al | Ta | Ti | Mo | C | B | Mg | Zr | Ni |
|---|---|---|---|---|---|---|---|---|---|---|---|---|
| VIM | 11–14 | 7–11 | 3–6 | 2–5 | 3–7 | 3–6 | 1–3 | 0.03–0.1 | trace | trace | trace | bal. |
| VIR | 11–13 | 8–10 | 4–5 | 2–5 | 3–6 | 3–5 | 1–3 | 0.06–0.1 | trace | trace | trace | bal. |

In the vacuum induction refining process, a variety of crucibles, including alumina crucibles, magnesium oxide crucibles, and zirconia crucibles, are commonly utilized for melting. Each type of crucible offers distinct properties and benefits, making them suitable for specific applications and alloy compositions. For the vacuum melting process discussed in this study, a magnesium oxide crucible was selected due to its particular advantages, which will be thoroughly characterized and detailed in Section 3.1.

### *2.2. Oxide Inclusion Characterization Methods*

Experimental samples for oxide inclusion characterization, cut from the core of the ignots, were prepared by electrodischarge machining. A field emission scanning electron microscope (FESEM, Crossbeam 550, Carl Zeiss AG, Oberkochen, Germany) equipped with an energy dispersive spectrometer (EDS, Ultim Max 65, Oxford, UK) was used to determine the oxide inclusion characteristics. The AZtecFeature automated particle analysis system, which incorporates a self-designed superalloy inclusion classification module, was used to quantitatively analyze the density, size distribution, and other characteristics of the oxide inclusions. The statistical area of the oxide inclusions in each sample was 2 mm × 1.5 mm, and oxide inclusions less than 300 nm in diameter were ignored in the statistical analysis.

It is worth noting that when statistically analyzing inclusions, we typically use a method that sequentially scans and collects data from multiple successive consecutive fields of view. In this experiment, we utilized a large field-of-view statistical technique specifically for analyzing oxide inclusions. This technique encompasses a substantial statistical area of 2 mm × 1.5 mm. The core principle of this method involves capturing 1000× magnification FESEM photographs, which are then concatenated to form a comprehensive and detailed overview. To enhance the accuracy and reliability of our statistical analysis, we took a total of 384 individual SEM photographs, covering the entire designated area. This extensive photographic dataset allows for a thorough and precise evaluation of the inclusions, ensuring that the data collected are both sufficient and representative.

In this work, the size of each oxide inclusion is represented by the equivalent circle diameter (ECD) with an equivalent plate area. According to the ECD results, the area fraction of the oxide inclusions in the experimental samples before and after the VIR process can be calculated. In order to further characterize the dispersion degree of the oxide inclusions in the samples, the test area was systematically divided into 300 smaller sections, each measuring 0.1 mm × 0.1 mm. This division allows for a detailed analysis of the distribution of oxide inclusions across the sample. The dispersion degree of the oxide inclusions is then quantified by calculating the standard deviation of the areal density of the oxide inclusions within each of these smaller areas. This statistical approach provides a comprehensive understanding of how uniformly the inclusions are distributed in the material before and after undergoing the VIR process.

## 3. Results

### *3.1. The Characteristics of the MgO Crucible Used in the VIR Process*

Figure 1 shows the morphology and phase analysis of the MgO crucible. As shown in Figure 1a, the MgO crucible is composed of two main phases. The phase analysis of a single field of view was not statistically significant. Therefore, the phase analysis of the MgO crucible is carried out using large-area mapping combined with AutoPhaseMap analysis. The phase analysis results are shown in Figure 1b,c, indicating that the MgO crucible is composed of 87.35% of the MgO phase and 12.65% of the $MgO{\cdot}Al_2O_3$ phase, in terms of volume fraction. The analysis results are consistent with other research findings, and the composition of the magnesium oxide melting crucible in the vacuum induction melting process consists of MgO and $MgO{\cdot}Al_2O_3$ phases.

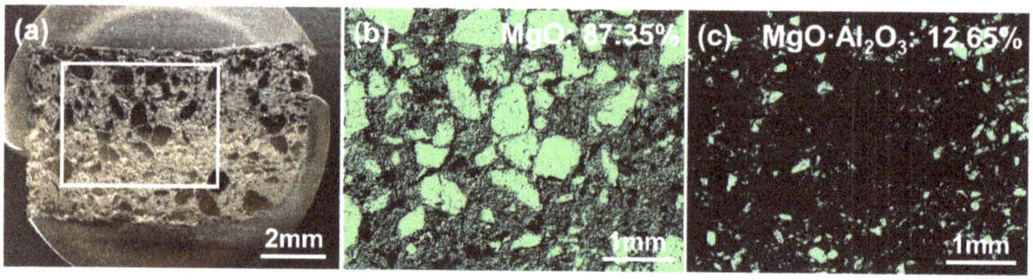


**Figure 1.** Phase analysis of the MgO crucible: (**a**) MgO crucible; a diagram of large-area mappings of (**b**) MgO and (**c**) $MgO{\cdot}Al_2O_3$ phase distributions.

In addition, the porous structure of the magnesium oxide crucibles can be clearly seen in Figure 1. Magnesium oxide crucibles are frequently prepared with a porous structure, primarily due to the high temperatures encountered during the vacuum melting process. This design choice is essential because dense structures tend to fracture under such conditions, often due to thermal expansion and other related stresses. The porous nature of these crucibles not only enhances their durability but also offers the additional benefit of being able to adsorb high melting point non-metallic or metallic oxides. This adsorption capability is crucial because it helps in trapping and retaining undesirable oxide inclusions, thereby improving the overall purity and quality of the molten metal.

### *3.2. Characteristics of Typical Oxide Inclusions before and after the VIR Process*

#### 3.2.1. The Morphology of Typical Oxide Inclusions

Figure 2 displays the morphology and EDS maps of the typical oxide inclusions found in the high-aluminum Ni-based superalloy before and after the VIR process. Figure 2a clearly shows that Mg, Al, and O are uniformly distributed throughout the inclusions, indicating that the typical oxide inclusions before the VIR process are made up of $MgO{\cdot}Al_2O_3$. In addition, many observational results show that most $MgO{\cdot}Al_2O_3$ inclusions are irregular in shape. Figure 2($b_1$–$b_4$) displays the morphologies and EDS mappings of typical oxide inclusions found in the Ni-based superalloy after the VIR process. In the remelted superalloy, endogenous $MgO{\cdot}Al_2O_3$, film-like $MgO{\cdot}Al_2O_3$, exogenous $MgO{\cdot}Al_2O_3$, and exogenous MgO inclusions are detected, as shown in Figure 2($b_1$–$b_4$), respectively. Figure 2($b_1$) clearly shows that Mg, Al, and O are uniformly distributed throughout the inclusions, indicating that the typical oxide inclusions are composed of endogenous $MgO{\cdot}Al_2O_3$. Figure 2($b_2$) shows that the film-like inclusions are composed of uniform Mg, Al, and O, indicating that the film-like inclusions are also $MgO{\cdot}Al_2O_3$. This may be because diluted air still exists in the vacuum during the VIR process, and the melt surface is rapidly oxidized. Therefore, when folded in or experiencing collisions between droplets, the surface oxide contacts the crucible liner when impinging against other masses of liquid [23]. This process may eventually lead to the formation of film-like $MgO{\cdot}Al_2O_3$. As shown in Figure 2($b_3$), the exogenous $MgO{\cdot}Al_2O_3$ inclusions are on the millimeter scale. This may be the result of the large MgO crucible matrix spalling into the superalloy melt and reacting with the $Al_2O_3$ inclusions or the dissolved [Al] in the superalloy melt. The exogenous presence of several tens of microns of MgO inclusions, as shown in Figure 2($b_4$), may also be caused by the spalling of the MgO crucible matrix.

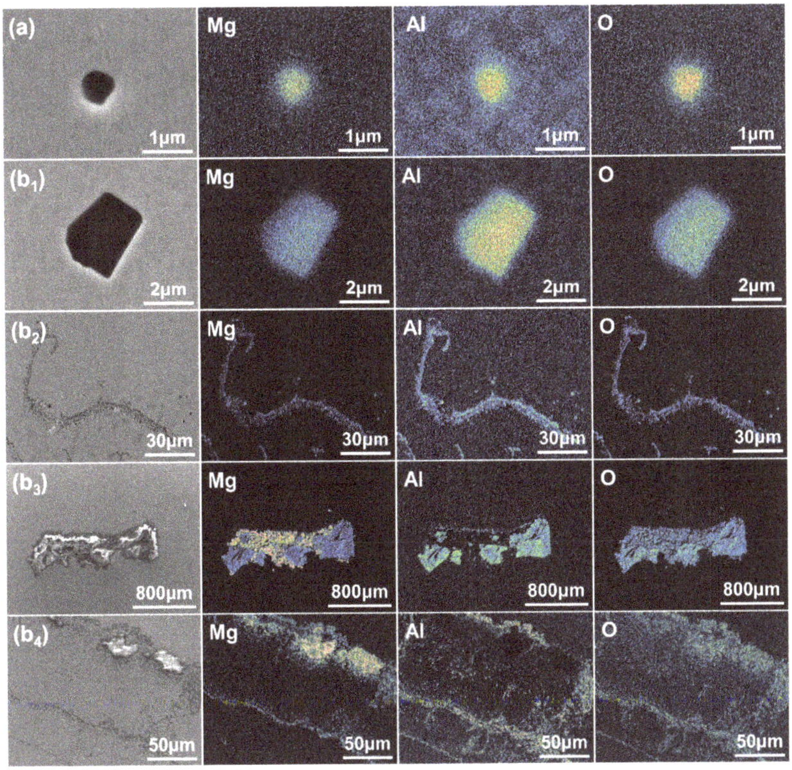


**Figure 2.** Morphologies and EDS mappings of typical oxide inclusions: (**a**) $MgO{\cdot}Al_2O_3$ inclusions found before the VIR process; (**$b_1$**) endogenous $MgO{\cdot}Al_2O_3$, (**$b_2$**) film-like $MgO{\cdot}Al_2O_3$, (**$b_3$**) exogenous $MgO{\cdot}Al_2O_3$, and (**$b_4$**) exogenous MgO inclusions found after the VIR process.

### 3.2.2. Types and Number Densities of Typical Oxide Inclusions

Table 2 summarizes the types and number densities of inclusions in the Ni-based superalloy before and after the VIR process. Based on the statistical results of the number of characteristics of oxide inclusions in high-aluminum Ni-based superalloys, the number densities of $MgO{\cdot}Al_2O_3$ before and after the VIR process are 62.67 and 45.00 $mm^{-2}$, respectively. It can be clearly seen that the VIR process can reduce the number density of $MgO{\cdot}Al_2O_3$ inclusions in the Ni-based superalloy. The decrease in the density of oxide inclusions is attributed to the porous material of the magnesium oxide crucible. As mentioned earlier, the magnesium oxide crucible used for vacuum melting is a porous material. During the VIR process, a portion of the oxide inclusions inside the alloy melt will be adsorbed inside the crucible, resulting in a decrease in the overall content of the oxides.

**Table 2.** The types and number densities of endogenous inclusions in high-aluminum Ni-based superalloys before and after the VIR process.

| Process | Type | Number | Area, mm | Number Density, $mm^{-2}$ |
|---|---|---|---|---|
| VIM | $MgO{\cdot}Al_2O_3$ | 188 | 3 | 62.67 |
| VIR | $MgO{\cdot}Al_2O_3$ | 145 | 3 | 45.00 |

### 3.2.3. The Size Distribution and Dispersion Degree of Typical Oxide Inclusions

The strength, toughness, and fatigue resistance of Ni-based superalloys are closely related to the size distribution and dispersion degree of oxide inclusions [24–26]. Therefore, describing these characteristics of oxide inclusions in Ni-based superalloys is crucial. The size of each oxide inclusion is represented by the equivalent circle diameter with an equivalent plate area. The dispersion degree is described by the standard deviation of the areal density of oxide inclusions for each small area.

Figure 3 shows the statistical results of the size distribution and dispersion degree of the oxide inclusions in the high-aluminum Ni-based superalloy before and after the VIR process. As shown in Figure 3a, the average $MgO{\cdot}Al_2O_3$ inclusion size before the VIR process is 0.64 μm. Most of the oxide inclusions, accounting for 99.49% of the total number of inclusions, are smaller than 2 μm. According to the ECD results, the area fraction of the inclusions in the sample before the VIR process is 0.00264%. Figure 3b reveals the average areal density of the oxide inclusions before the VIR process, and the standard deviation is 1.180, indicating that the sizes of the $MgO{\cdot}Al_2O_3$ inclusions before the VIR process are not relatively uniform. As shown in Figure 3c, the average oxide inclusion size after the VIR process is 0.76 μm. Most of the oxide inclusions, accounting for 95.10% of the total number of inclusions, are smaller than 2 μm. Compared with the average size of oxide inclusions before the VIR process, the size of the inclusions increases, which is the inevitable result of the aggregation and growth of oxide inclusions during the VIR process. According to the ECD results, the area fraction of inclusions in the sample after the VIR process is 0.00335%. Figure 3d reveals the average areal density of the oxide inclusions after the VIR process, and the standard deviation is 0.737, indicating that the sizes of the $MgO{\cdot}Al_2O_3$ inclusions after the VIR process are more uniform than those before the VIR process. Therefore, the statistical results show that the VIR process can increase the oxide inclusion size and dispersion degree of inclusions to a certain extent in high-aluminum Ni-based superalloys.

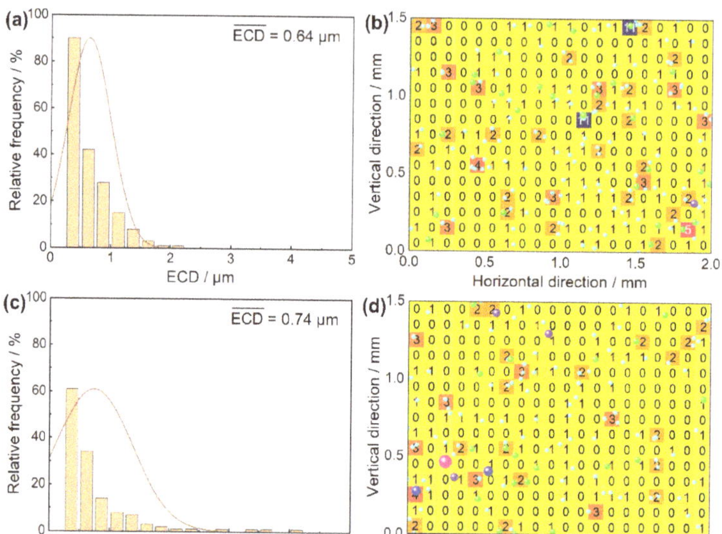


**Figure 3.** The statistical results of endogenous $MgO{\cdot}Al_2O_3$ inclusions: (**a**) the size distribution (**b**) and areal density distribution before the VIR process, (**c**) and the size distribution and (**d**) areal density distribution after the VIR process. (cyan ball: <1 μm, green ball: 1–2 μm, purple ball: 2–3 μm, blue ball: 3–4 μm, pink ball: 4–5 μm).

## 4. Discussion

### 4.1. The Formation Mechanism of Endogenous Irregular $MgO{\cdot}Al_2O_3$ Inclusions after the VIR Process

The compositions of the typical small inclusions in the experimental samples before and after the VIR process at room temperature are consistent with each other, indicating that the endogenous $MgO{\cdot}Al_2O_3$ inclusions after the VIR process are mainly derived from the master alloy. To verify our hypothesis, the phase transformation and oxide precipitation behavior before and after the addition of Mg to the molten alloy during the VIM process were calculated using FactSage 8.3 software.

The thermodynamic equilibrium calculations before and after the addition of Mg to the molten alloy are shown in Figure 4a,b, respectively. As shown in Figure 4a, the oxide inclusions before Mg's addition only existed in the form of the CORU#1 phase above 1573 K. The CORU#1 phase consisted of $Al_2O_3$ containing trace amounts of $Ti_2O_3$, $Cr_2O_3$, $Fe_2O_3$, and $Mn_2O_3$. However, after the addition of Mg, as shown in Figure 4b, oxide inclusions only existed in the form of the TiSp phase above 1583 K. The TiSp phase is $MgAl_2O_4$ and contains trace amounts of $MgTi_2O_4$, $MnTi_2O_4$, and $FeTi_2O_4$. The above experimental results show that the typical oxide inclusion before the VIR process is $MgO{\cdot}Al_2O_3$, which is consistent with the calculation results. Notably, the typical oxide inclusion after the VIR process is also $MgO{\cdot}Al_2O_3$, indicating that the VIR process has no significant influence on the composition of inclusions in high-aluminum Ni-based superalloys.

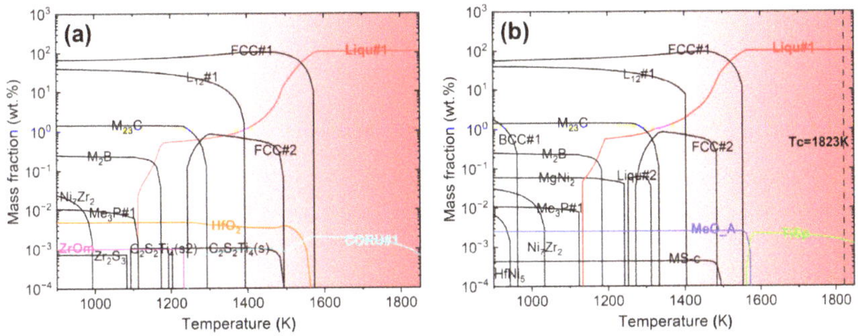


**Figure 4.** The thermodynamic equilibrium calculation results (**a**) before Mg's addition and (**b**) after Mg's addition during solidification.

### 4.2. The Formation Mechanism of Film-like $MgO{\cdot}Al_2O_3$ Inclusions after the VIR Process

As shown in Figure 5, SEM and EDS mapping were performed to investigate the interface morphologies and distributions of the main elements between the MgO crucible and the high-aluminum Ni-based superalloy after the VIR process. As shown in Figure 5a, b, several hundred microns of massive and elongated inclusions are easily observed. In addition, there is a dark reaction layer on the MgO crucible side and the superalloy side. To determine the composition of the two dark reaction layers, EDS mapping analysis was performed on the regions. Figure 5c clearly shows that the interface is mainly composed of a $MgO{\cdot}Al_2O_3$ layer and an $Al_2O_3$ layer near the MgO crucible side. The $MgO{\cdot}Al_2O_3$ layer is much thinner than the $Al_2O_3$ layer and has a thickness of approximately 3.60 μm compared to 19.94 μm. The morphology and EDS mappings near the superalloy side are shown in Figure 5d. The interface is mainly composed of an $Al_2O_3$ inclusion layer near the superalloy side at approximately 1.51 μm. Large, film-like $MgO{\cdot}Al_2O_3$ and $Al_2O_3$ inclusions are also detected at the interface between the MgO crucible and the superalloy. This may be due to the adsorption of $Al_2O_3$ inclusions in the melt by the inner wall of the MgO crucible and the subsequent solid–solid reaction to form film-like $MgO{\cdot}Al_2O_3$, which is ultimately involved in the superalloy melt.

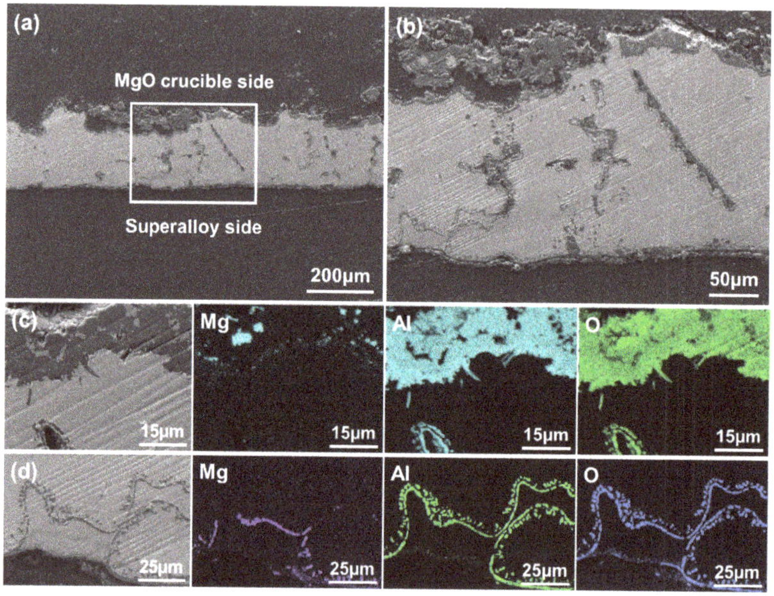


**Figure 5.** The morphologies and EDS maps of the interface between the MgO crucible and the high-aluminum Ni-based superalloy after the VIR process. (**a**) Low-magnification SEM image; (**b**) high-magnification SEM image; (**c**) morphologies and EDS mappings near the MgO crucible side; (**d**) morphologies and EDS mappings near the superalloy side.

To verify the above hypothesis, thermodynamic calculations were performed to explore the effect of the crucible and alloy melt interface reaction during the VIR process on the formation of oxide inclusions in the final superalloy components. The thermodynamic calculations of the interfacial reaction between the MgO crucible and the high-aluminum Ni-based superalloy at 1823 K are shown in Figure 6. The MeO_A phase consists of MgO containing trace amounts of $Al_2O_3$, $Cr_2O_3$, $ZrO_2$, etc., while the TiSp phase consists of $MgAl_2O_4$ containing trace amounts of $MgTi_2O_4$, $MnTi_2O_4$, $FeTi_2O_4$, etc. It can be clearly seen that a large amount of the TiSp phase appears at the interface near the alloy melt side. As the distance from the melt side increases, the content of the TiSp phase increases. The calculated results are consistent with the experimental results, demonstrating that the interfacial reaction between the crucible and superalloy melt during the VIR process introduces many $MgO \cdot Al_2O_3$ inclusions into the interface.

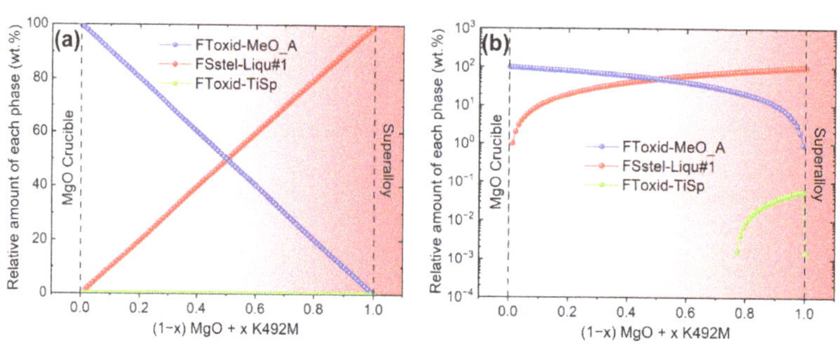


**Figure 6.** The thermodynamic calculation results of the interfacial reaction between the MgO crucible and the high-aluminum Ni-based superalloy at 1823 K: (**a**) the linear coordinate system and (**b**) the logarithmic coordinate system.

During the VIR process, $Al_2O_3$ inclusions may exist in the superalloy melt due to the so-called vacuum, which must be viewed as merely diluted air. The dissolved [Al] in a high-aluminum Ni-based superalloy melt can easily react with oxygen in diluted air to form $Al_2O_3$ inclusions. To further determine the mechanism responsible for the formation of $MgO \cdot Al_2O_3$ inclusions between the MgO crucible and the high-aluminum Ni-based superalloy after the VIR process, the Al-MgO and $Al_2O_3$-MgO phase diagrams were calculated, as shown in Figure 7a,b, respectively. Clearly, Al cannot directly react with MgO, while $Al_2O_3$ can react with MgO to form $MgAl_2O_4$. At the casting temperature, the reaction between MgO and $Al_2O_3$ is efficient, requiring only 0.10% $Al_2O_3$. Therefore, the $MgO \cdot Al_2O_3$ inclusions at the interface are the result of an in situ solid–solid reaction between the MgO crucible and the $Al_2O_3$ inclusions in the superalloy melt.

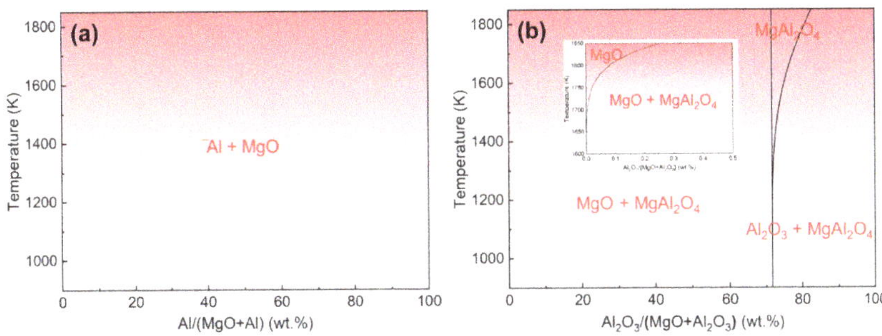


**Figure 7.** Binary phase diagrams: (**a**) Al-MgO and (**b**) $Al_2O_3$-MgO.

### *4.3. The Formation Mechanism of Large $MgO \cdot Al_2O_3$ and MgO Inclusions after the VIR Process*

Overall, the formation mechanism of $MgO \cdot Al_2O_3$ inclusions at the interface can be summarized as follows: (i) the dissolved [Al] in the superalloy melt reacts with $O_2$ to form $Al_2O_3$ inclusions; (ii) $Al_2O_3$ inclusions randomly adhere to the inner wall of the MgO crucible; (iii) an in situ solid–solid phase reaction occurs between the $Al_2O_3$ and MgO at the interface to form $MgO \cdot Al_2O_3$ inclusions; and (iv) the $MgO \cdot Al_2O_3$ inclusions that form at the interface peel off and entrap into the superalloy melt. The detailed formation process of $MgO \cdot Al_2O_3$ inclusions in the high-aluminum Ni-based superalloy during the VIR process is illustrated in Figure 8, which provides a visual representation of each step in the mechanism. This understanding is crucial for developing strategies to control and minimize the presence of such inclusions in high-aluminum nickel-based superalloys.

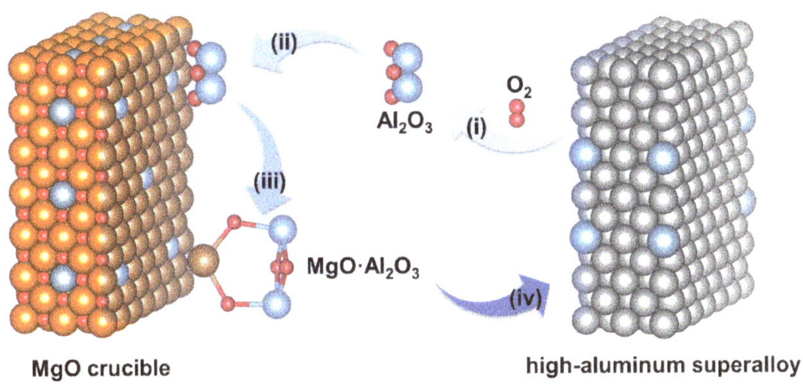


**Figure 8.** The formation mechanism of large $MgO \cdot Al_2O_3$ inclusions in the high-aluminum Ni-based superalloy during the VIR process.

## 5. Conclusions

The evolution and formation mechanism of oxide inclusions in a high-aluminum Ni-based superalloy were quantitatively investigated through a combination of experiments and thermodynamic calculations during the VIR process. The main conclusions are summarized as follows:

(1) The typical oxide inclusions in the master alloy and high-aluminum Ni-based superalloy after the VIR process are both made up of irregular $MgO \cdot Al_2O_3$. The VIR process effectively reduced the number density of the oxide inclusions in the high-aluminum Ni-based superalloy from 62.67 to 45.00 $mm^{-2}$. Simultaneously, the average oxide inclusion size increases from 0.64 to 0.74 μm, the area fraction of inclusions increases from 0.00264 to 0.00335%, and the standard deviation of the areal density of the oxide inclusions decreases from 1.180 to 0.737, indicating an increase in oxide inclusion uniformity.

(2) While the VIR process demonstrates efficacy in reducing oxide inclusions, it introduces millimeter-scale oxide inclusions, several tens of microns of MgO inclusions, and large micron-scale film inclusions into high-aluminum Ni-based superalloys. It is imperative to rigorously eliminate these introduced inclusions. Consequently, selecting a high-quality crucible during the smelting process is of paramount importance.

(3) The VIR process in high-aluminum Ni-based superalloys involves three primary formation mechanisms for $MgO \cdot Al_2O_3$. These mechanisms include the heredity of oxide inclusions, the interfacial reaction between the MgO crucible and the high-aluminum superalloy melt, and the exfoliation and entrapment of the inner wall of the MgO crucible, which adsorbs $Al_2O_3$ inclusions and undergoes chemical reactions. The second three mechanisms introduce several hundred micron-thick film-like and bulk millimeter-scale $MgO \cdot Al_2O_3$ inclusions.

**Author Contributions:** Conceptualization, Y.Z. and M.X.; methodology, M.X.; software, E.L. and W.F.; validation, W.X., Z.X. and W.F.; formal analysis, L.Z., W.X. and C.G.; investigation, Y.L.; resources, C.G.; data curation, Y.L.; writing—original draft preparation, E.L.; writing—review and editing, E.L. and W.X.; visualization, Z.X.; supervision, M.X.; project administration, L.Z. and Y.Z.; funding acquisition, L.Z. and Y.Z. All authors have read and agreed to the published version of the manuscript.

**Funding:** This research was supported by the National Natural Science Foundation of China (No. 52374360).

**Data Availability Statement:** The raw data supporting the conclusions of this article will be made available by the authors on request.

**Conflicts of Interest:** Author Lihui Zhang, Weijie Xing, Yunsong Zhao, Yushi Luo was employed by the company Beijing Institute of Aeronautical Materials. The remaining authors declare that the research was conducted in the absence of any commercial or financial relationships that could be construed as a potential conflict of interest.

## References

1. Yang, C.; Hu, R.; Wang, X.; Du, J.; Luo, X.; Bi, Z.; Gan, B. Effect of pre-tensile treatments on the mechanical properties and deformation mechanism of a novel Ni-based superalloy. *Mater. Sci. Eng. A* **2023**, *874*, 145063. [CrossRef]
2. Juillet, C.; Oudriss, A.; Balmain, J.; Feaugas, X.; Pedraza, F. Characterization and oxidation resistance of additive manufactured and forged IN718 Ni-based superalloys. *Corros. Sci.* **2018**, *142*, 266–276. [CrossRef]
3. Detrois, M.; Antonov, S.; Rozman, K.A.; Hawk, J.A.; Jablonski, P.D. Improved creep and tensile properties of a corrosion resistant Ni-Based superalloy using high temperature aging and Nb/Ta additions. *Metall. Mater. Trans. A* **2022**, *53*, 2600–2613. [CrossRef]
4. Wang, J.; Wang, L.; Li, J.; Chen, C.; Yang, S.; Li, X. Effects of aluminum and titanium additions on the formation of nonmetallic inclusions in nickel-based superalloys. *J. Alloys Compd.* **2022**, *906*, 164281. [CrossRef]
5. Bergsmo, A.; Xu, Y.; Poole, B.; Dunne, F.P.E. Twin boundary fatigue crack nucleation in a polycrystalline Nickel superalloy containing non-metallic inclusions. *J. Mech. Phys. Solids* **2022**, *160*, 104785. [CrossRef]
6. Yao, Z.; Hou, J.; Chen, Y.; Xu, W.; Jiang, H.; Dong, J. Effect of micron-sized particles on the crack growth behavior of a Ni-based powder metallurgy superalloy. *Mater. Sci. Eng. A* **2022**, *860*, 144242. [CrossRef]

7. Liu, P.; Jiang, H.; Dong, J.; Chen, Z. Effect of micron-scale nonmetallic inclusions on fatigue crack nucleation in a nickel-based superalloy. *Int. J. Solids Struct.* **2023**, *279*, 112368. [CrossRef]
8. Texier, D.; Stinville, J.C.; Echlin, M.L.P.; Pierret, S.; Villechaise, P.; Pollock, T.M.; Cormier, J. Short crack propagation from cracked non-metallic inclusions in a Ni-based polycrystalline superalloy. *Acta Mater.* **2019**, *165*, 241–258. [CrossRef]
9. Zhao, Y.; Wang, L.; Chen, C.; Li, J.; Li, X. Effect of a MgO–CaO–$ZrO_2$-based refractory on the cleanliness of a K4169 Ni-based superalloy. *Ceram. Int.* **2023**, *49*, 117–125. [CrossRef]
10. Gao, R.; Li, L.; Chen, C.; Li, J.; Wang, L.; Li, X.; Yang, H. Formation and aggregation behavior of inclusions in Ni-based alloys with different Mg contents. *J. Mater. Res. Technol.* **2023**, *26*, 5252–5263. [CrossRef]
11. Miller, C.F.; Simmons, G.W.; Wei, R.P. Mechanism for oxygen enhanced crack growth in inconel 718. *Scr. Mater.* **2001**, *44*, 2405–2410. [CrossRef]
12. Jiang, R.; Ji, D.; Shi, H.; Hu, X.; Song, Y.; Gan, B. Effects of thermal exposure on high-cycle-fatigue behaviours in Ni-based superalloy GH4169. *Mater. Sci. Technol.* **2019**, *35*, 1265–1274. [CrossRef]
13. Yang, S.; Yang, S.; Qu, J.; Du, J.; Gu, Y.; Zhao, P.; Wang, N. Inclusions in wrought superalloys: A review. *J. Iron Steel Res. Int.* **2021**, *28*, 921–937. [CrossRef]
14. Chan, K.S. A fatigue life model for predicting crack nucleation at inclusions in Ni-based superalloys. *Metall. Mater. Trans. A* **2020**, *51*, 1148–1162. [CrossRef]
15. Akande, I.G.; Oluwole, O.O.; Fayomi, O.S.I.; Odunlami, O.A. Overview of mechanical, microstructural, oxidation properties and high-temperature applications of superalloys. *Mater. Today Proc.* **2021**, *43*, 2222–2231. [CrossRef]
16. Gao, X.; Zhang, L.; Qu, X.; Luan, Y.; Chen, X. Investigation on the formation mechanism of non-metallic inclusions in high-aluminum and titanium-alloyed Ni-based superalloy. *Vacuum* **2020**, *177*, 109409. [CrossRef]
17. Liu, Y.; Gao, Y.; Wang, E.; Chen, G.; Xu, E.; Zhao, F.; Zhao, Y.; Li, C.; Hou, X. Interaction between $CA_6$-MA crucible and molten wrought Ni-based superalloys. *J. Eur. Ceram. Soc.* **2023**, *43*, 1714–1722. [CrossRef]
18. Song, Q.; Qian, K.; Shu, L.; Chen, B.; Ma, Y.; Liu, K. Interfacial reaction between nickel-based superalloy K417G and oxide refractories. *Acta Metall. Sin.* **2021**, *58*, 868–882.
19. Gao, X.; Zhang, L.; Qu, X.; Chen, X.; Luan, Y. Effect of interaction of refractories with Ni-based superalloy on inclusions during vacuum induction melting. *Int. J. Miner. Metall. Mater.* **2020**, *27*, 1551–1559. [CrossRef]
20. Li, J.; Zhang, H.; Gao, M.; Li, Q.; Liu, H.; Zhang, H. Effect of vacuum level on the interfacial reactions between K417 superalloy and $Y_2O_3$ crucibles. *Vacuum* **2020**, *182*, 109701. [CrossRef]
21. Yiliti, Y.; Dong, G.; Liu, X.; You, X.; Han, W.; Dong, L.; Zhao, Y.; Yi, L.; Wang, Y. The high temperature oxidation behavior of a superalloy prepared by vacuum induction melting and electron beam smelting: A comparative study. *J. Mater. Res. Technol.* **2023**, *25*, 6977–6991. [CrossRef]
22. You, X.; Dong, G.; Zhou, H.; Zhang, H.; Tan, Y.; Wang, Y.; Li, P.; You, Q.; Li, Y.; Cui, H.; et al. Removal of oxygen, nitrogen, and inclusions in powder superalloy scraps by electron beam smelting and induced solidification and the purification mechanisms. *Sep. Purif. Technol.* **2023**, *304*, 122290. [CrossRef]
23. Campbell, J.; Tiryakioğlu, M. Bifilm defects in Ni-based alloy castings. *Metall. Mater. Trans. B* **2012**, *43*, 902–914. [CrossRef]
24. Jiang, J.; Yang, J.; Zhang, T.; Dunne, F.P.E.; Britton, T.B. On the mechanistic basis of fatigue crack nucleation in Ni superalloy containing inclusions using high resolution electron backscatter diffraction. *Acta Mater.* **2015**, *97*, 367–379. [CrossRef]
25. Hu, D.; Wang, T.; Ma, Q.; Liu, X.; Shang, L.; Li, D.; Pan, J.; Wang, R. Effect of inclusions on low cycle fatigue lifetime in a powder metallurgy nickel-based superalloy FGH96. *Int. J. Fatigue* **2019**, *118*, 237–248. [CrossRef]
26. Zheng, L.; Zhang, G.; Gorley, M.J.; Lee, T.L.; Li, Z.; Xiao, C.; Tang, C.C. Effects of vacuum on gas content, oxide inclusions and mechanical properties of Ni-based superalloy using electron beam button and synchrotron diffraction. *Mater. Des.* **2021**, *207*, 109861. [CrossRef]

metals

MDPI

Article

# Strip Casting of $Sm_2TM_{17}$-Type Alloys for Production of the Metastable $SmTM_7$ Phase

**Richard Sheridan *, Joseph Gresle-Farthing, Alice Appleby and Mangaliso Brown**

School of Metallurgy and Materials, University of Birmingham, Edgbaston, Birmingham B15 2TT, UK; j.greslefarthing@bham.ac.uk (J.G.-F.); axa1291@bham.ac.uk (A.A.); m.brown.4@bham.ac.uk (M.B.)

* Correspondence: r.s.sheridan.1@bham.ac.uk

**Abstract:** Conventional book casting of $Sm_2TM_{17}$-type alloys (where TM = Co, Fe, Cu, Zr) leads to a coarse, highly segregated microstructure, predominantly due to the slow, variable cooling rate from the mould surface towards the centre of the ingot. These cast alloys require a long homogenisation treatment to remove this segregation and develop a super-saturated, metastable $SmTM_7$-type hexagonal phase. This $SmTM_7$ phase is a vital precursor phase required during magnet production to develop the complex cellular structure responsible for high magnetic properties. In this work, strip casting was employed to facilitate rapid solidification to develop thin flakes (<0.5 mm thick) with a columnar grain structure. Rapid cooling has the potential to produce a homogenous microstructure consisting predominantly of the metastable $SmTM_7$ phase. This could remove or significantly reduce the need for the energy-intensive homogenisation treatment usually required in conventional magnet manufacture. This paper investigates the effect of wheel speed (and hence cooling rate) on flake thickness, microstructure, and phase balance of the cast alloys. It was shown that for wheel speeds between 1.1 and 3.0 m/s, the microstructure showed large variation; however, in all cases, evidence of the columnar $SmTM_7$ phase was presented. The adhesion between the melt and the wheel was deemed to be critical for the nucleation and subsequent columnar growth of $SmTM_7$ grains, where the wheel speed controlled both the flow of the alloy onto the wheel and the thickness of the resultant flake. It was determined that in order to achieve a homogenous columnar $SmTM_7$ structure, the maximum flake thickness should be limited to 270 μm to avoid the formation of equiaxed $Sm_2TM_{17}$ grains through insufficient cooling.

**Keywords:** samarium cobalt; strip casting; metastable; rapid solidification

## 1. Introduction

Samarium cobalt (SmCo) magnets are high energy density permanent magnets suitable for applications such as high-speed motors and capable of operating at temperatures up to 550 °C. This is due to their excellent magnetic thermal coefficients, high Curie temperature, and a protective oxide ($Sm_2O_3$) layer, making them ideal for extreme environments [1]. One type of SmCo magnet is the 2:17 type, often referred to as $Sm_2TM_{17}$, where TM stands for transition metals (Co, Fe, Cu, Zr). The name 2:17 refers to the atomic ratio of Sm to transition metals in the hard-magnetic matrix phase. This nomenclature, as well as a capital letter denoting the crystal structure, will be used to refer to phases throughout this paper (see Table 1).

$Sm_2TM_{17}$ magnets are highly dependent on a very particular phase structure, chemical composition, and heat treatment regime utilised during the manufacturing process [2]. This is a sub-granular nano-cellular microstructure that is comprised of FeCo-rich 2:17R rhombohedral $Th_2Ni_{17}$-type phase cells, providing the high energy product and remanence of the magnet [3]. These are surrounded by a Cu-rich boundary 1:5H hexagonal phase. The Cu concentration gradient between the cells and the boundary phase provides domain wall pinning, which is the dominant coercivity mechanism in 2:17 magnets [4]. A Zr-rich phase forms a superimposed lamella and acts as a diffusion pathway during processing, allowing enrichment of the 2:17R cells with Fe and Co and the 1:5H boundary phase with Cu [5]. To form this structure, $Sm_2TM_{17}$ magnets undergo long, high temperature heat treatments: firstly, a solution heat treatment at 1100–1200 °C for 4–10 h to form a metastable phase (1:7H), which is stabilised at room temperature by rapid quenching. This phase is described as disordered 2:17 in some literature and adopts a $TbCu_7$ structure [6,7] which decomposes during subsequent heat treatment to develop a uniform, optimum micro/nano-structure [8–10]. This is followed by ageing at 750–850 °C for up to 12 h which allows the precipitation of the fine 2:17R cells and the formation and enrichment of the boundary phase from the precursor 1:7H phase. Further ageing is achieved by slow cooling to 400 °C and holding for 2–10 h to finesse the nanostructure of the magnet [11]. It should be noted that times and temperatures vary between manufacturers depending on specific elemental composition, desired grain size, desired cellular structure, and target magnetic properties.

**Table 1.** Reference table for the terminology used in this paper to describe the phases present within the microstructure.

| Phase | Sm:TM Ratio | Crystal Structure | Crystal Structure Type |
|---|---|---|---|
| 1:7H | 1:7 | Hexagonal | $TbCu_7$ |
| 1:5H | 1:5 | Hexagonal | $CaCu_5$ |
| 2:17R | 2:17 | Rhombohedral | $Th_2Ni_{17}$ |
| 2:17H | 2:17 | Hexagonal | $Th_2Zn_{17}$ |

In traditional manufacturing routes, the alloy is cast into water-cooled steel/copper moulds, causing variable cooling rate during solidification. The edges experience an extremely high cooling rate which diminishes towards the centre of the ingot. The variation in cooling rate causes an extremely inhomogeneous microstructure, segregated with large fractions of undesirable phases. Furthermore, these alloys are difficult to process, requiring high-energy jaw crushing and milling to reduce the particle size significantly. The composition and microstructure remain inhomogeneous, and hence heat treatment is crucial to achieving the desired magnetic properties [12]. The macrosegregation within the alloy is characterised by large regions rich in Fe and Co, with the other regions becoming enriched with Sm and Cu. Small amounts of the desirable metastable 1:7H phase are present in cast ingots, at the edges where the cooling rate is highest [13]. Theoretically, if the material could be cast with a sufficiently high cooling rate, the homogenisation step of the heat treatment could be skipped to save energy and thus cost.

Strip casting is a technique used in the production of other rare earth magnets that allows for rapid non-equilibrium cooling, potentially producing more desirable microstructures [14–17]. Strip casting works by pouring the melt onto a spinning water-cooled copper wheel, allowing for rapid solidification at cooling rates of around $10^4$ K/s and producing long, thin flakes of the alloy. The wheel rotation speed is directly proportional to the cooling rate of the alloy but inversely proportional to the thickness of the flake it produces. The flake thickness of the alloy is typically between 100 and 1000 μm compared to ingots, which are typically >10 mm [13,17]. Strip cast flakes generally have a microstructure that varies along the direction of solidification across the thickness of the flake. Where the alloy initially contacts the wheel, a nucleation zone of rapidly solidified alloy forms. From this zone, directional columnar growth begins, consisting of fine needle-like grains with widths of only a few microns. As they grow along the direction of cooling towards the free side, eventually the rate of cooling is not sufficient to sustain their directional growth, and they lose their orientation, with their tips acting as nucleation sites for more randomly oriented grains. These grains are still relatively fine; however, they coarsen towards the free side of the strip. Long, thin flakes could also give a significant advantage during crushing and milling, potentially requiring lower energy techniques to reduce particle size for magnet manufacture due to the highly brittle nature of the flakes.

Strip casting of $Sm_2TM_{17}$ has only been explored in a few published papers, with only one paper seeking to obtain the 1:7H phase directly. Liu et al. found that casting at a wheel speed of 5 m/s produced a homogenous 1:7H phase structure with superficial nanoscale grains [18]. However, this was performed using a melt spinner, which is not truly representative of strip casting. The melt is ejected from the crucible onto the wheel using high pressure gas, giving cooling rates up to $10^6$ K/s and producing ribbons that are typically 20–50 μm thick, compared to strip casting, where the alloy is gravity fed onto the wheel, leading to higher thicknesses. This work presented by Liu et al. led to non-directional growth as the cooling rate was too high to allow for columnar, directional growth along a cooling gradient, which is synonymous with strip casting [18]. Meng et al. also investigated strip casting of $Sm(CoFeCuZr)_z$ alloys for magnet manufacture but did not detail the strip casting method or parameters used [19]. They found that the fine grains developed formed the majority of the strip cast alloy and were very detrimental to the remanence and energy product of sintered magnets. When the strip cast alloy was milled, the grain size in these regions was smaller than the average particle size, which led to polycrystalline particles with poor alignment in the final magnet. Therefore, an alloy consisting of columnar grains would likely be more beneficial to magnet production than the fine, equiaxed grains produced in this work [19]. Yang et al. [20] investigated the effect of strip casting wheel speed as a part of a larger study on magnet manufacture. They found that slower wheel speeds were more conducive to forming a microstructure suitable for magnet manufacture; however, they were not aiming for direct production of the 1:7H phase and were completing the entire manufacturing route as would be used for cast ingots containing heavy elemental segregation. Whilst this showed very promising results from a magnetic point of view, it did not focus on direct production of the metastable 1:7H phase [20]. Yang et al. [21] investigated hydrogen decrepitation of strip cast alloys similar to those presented in [20]. However, the micrograph showing the full cross-section of the alloy showed a rough surface that solidified in contact with the wheel, with a small nucleation zone and predominantly non-directional or equiaxed growth. There was very little evidence of the characteristic columnar growth associated with strip casting. Zheng et al. [22] also investigated the hydrogen absorption of strip cast $Sm_2TM_{17}$-type alloys; however, in this work, there was no attempt to optimise the strip casting process. The authors produced ~300 μm thick flakes with some columnar growth, although there was a large amount of variation in the directionality of the columnar grains. However, they did report up to ~30–32 wt% 1:7H phase in their alloys, which shows great promise for the work presented in this paper [22].

These studies have shown that non-equilibrium casting has the potential for tailoring the microstructure to avoid the need for long term heat treatment. This project instead aims to investigate the viability of strip casting $Sm_2TM_{17}$-type alloys to circumvent energy intensive homogenisation heat treatment in the production of sintered magnets.

## 2. Materials and Methods

In this work, 4.5 kg of master ingots of the composition $Sm_{26.6}Co_{Bal}Fe_{19.9}Cu_{4.4}Zr_{2.3}$ (wt%) were loaded into the induction furnace of the strip caster for each individual casting. The alloy was heated to 1380 °C under a partial pressure of argon (200 mbar). The water-cooled wheel was ground with P120 SiC grinding paper to provide surface texture for the alloy to grip the wheel during casting. All casting parameters were kept constant, with the exception of the wheel rotational speed, which was varied across three casting runs with the following wheel speeds: Run 1 was cast at 1.1 m/s, Run 2 was cast at 2.1 m/s, and Run 3 was cast at 3.0 m/s. A schematic representation of the strip caster is shown in Figure 1.

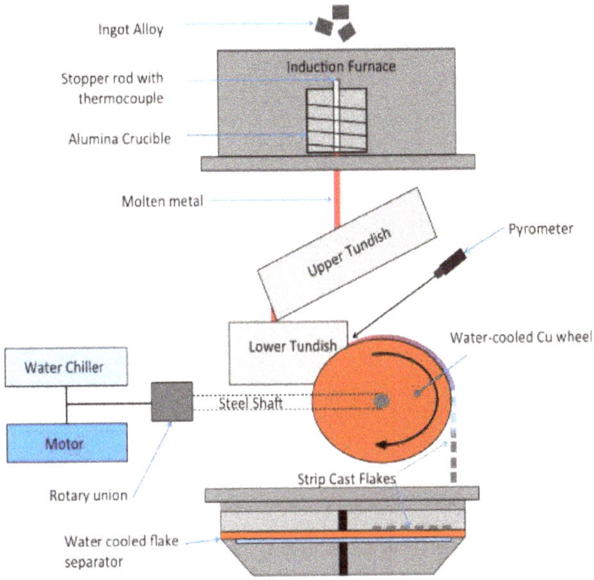


**Figure 1.** Labelled schematic of a pilot-scale strip caster at the University of Birmingham showing key components and flow of molten alloy from an induction furnace through a tundish system onto a water-cooled wheel to produce strip cast flakes, Adapted from Ref. [17].

Strip cast flake thickness measurements were taken using a digital micrometer, taking 100 measurements of randomly selected flakes to calculate the average flake thickness. The flakes were then broadly divided into three categories depending on thickness, and samples from each thickness range were mounted in cold setting epoxy resin for microscopy. These samples were ground using P800, P1200, P2400, and P4000 SiC grinding papers for 5 min per grit using an oil-based lubricant to prevent any reaction with water. Samples were subsequently polished using a 6 μm, 1 μm, and 0.25 μm diamond compound in an oil based suspension for 10 min each using a Selvyt cloth. Scanning electron microscopy (SEM) was performed using a Hitachi 4000TMplus microscope (Tokyo, Japan) with an energy dispersive X-ray (EDX) detector in backscatter electron (BSE) mode with an accelerating voltage of 15 kV, a working distance of 9–11 mm, and magnification chosen to image the cross-section of the entire flake thickness for each sample. X-ray diffraction (XRD) was performed using Proto XRD with a 2θ range of 20–90° and a dwell time of 1 s.

## 3. Results

### *3.1. As-Received Ingot*

The as-received ingot was provided by Less Common Metals Ltd., Ellesmere Port, UK. The microstructure is shown in Figure 2 and demonstrates large areas of elemental segregation. The lighter regions (labelled B) were found to be rich in Sm and Cu, and the darker regions (labelled A) were rich in Co and Fe. XRD analysis identifies that the microstructure is predominantly the 2:17H phase with small amounts of 2:17R, 1:5H, and 1:7H phases present.

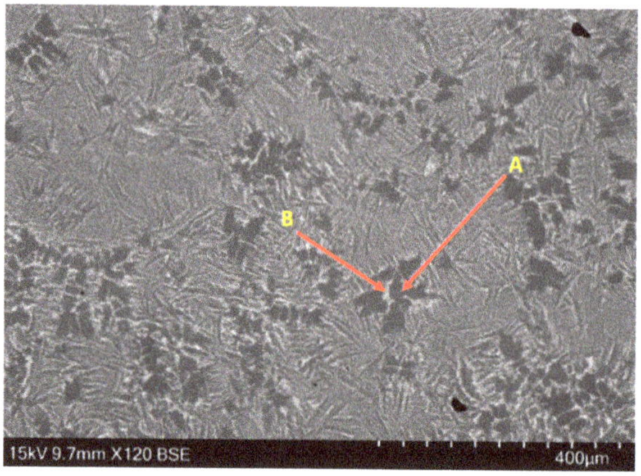


**Figure 2.** Microstructure of as-received ingot showing heavily segregated regions where label A represents a Co/Fe-rich phase and label B represents a Sm/Cu-rich phase.

### 3.2. Strip Cast Flakes

Thickness measurements for the three casting runs are shown in Figure 3. It was expected that as wheel speed increased, the flake thickness would decrease. This is because at higher wheel speeds, the alloy would be drawn onto the wheel more quickly and spread into thinner sheets. The mean flake thickness (and standard deviation) for Run 1 was 0.43 mm (0.06 mm), Run 2 was 0.45 mm (0.08 mm), and Run 3 was 0.40 mm (0.08 mm).

The microstructure of flakes taken from a variety of flake thicknesses from Runs 1–3 are shown in Figure 4. Each micrograph shows the entire cross-section of the flake, where the side directly in contact with the wheel during solidification is on the left and the top, free surface is on the right. The cooling direction is marked by a blue arrow. Each micrograph is accompanied by a coloured bar; the blue section indicates the nucleation zone where matrix grains are nucleated and columnar growth originates. The ideal microstructure would contain a short, uniform nucleation zone. The red zone indicates the region of columnar growth; this is where the cooling rate is optimal and is the ideal microstructure for further processing. The yellow section indicates the region where the cooling rate is too slow, columnar growth is no longer promoted, and non-directional grain growth occurs, along with elemental segregation, which requires heat treatment to remove. Quantification of each of these zones is presented in Table 2, where ranges of values are taken for each section as they are not completely uniform throughout the entire cross-section of the sample.

**Table 2.** Measurements of each microstructural zone within the samples selected from Runs 1, 2, and 3.

| Sample | Nucleation Zone (μm) | Columnar Zone (μm) | Non-Direction Growth (μm) | Flake Thickness (μm) |
|---|---|---|---|---|
| Run 1 Thin | 16 | 222–235 | 3–25 | >252 |
| Run 1 Median | Not Visible | 102–140 | 245–267 | >410 |
| Run 1 Thick | 17 | 112–179 | 236–275 | >416 |
| Run 2 Thin | 19 | 225–251 | 26–46 | >300 |
| Run 2 Median | 20 | 178–207 | 248–283 | >457 |
| Run 2 Thick | 16 | 201 | 197 | >418 |
| Run 3 Thin | 11 | 70–112 | 73–101 | >201 |
| Run 3 Median | 7 | 53–90 | 216–280 | >312 |
| Run 3 Thick | Not Visible | 52–68 | 311–350 | >417 |

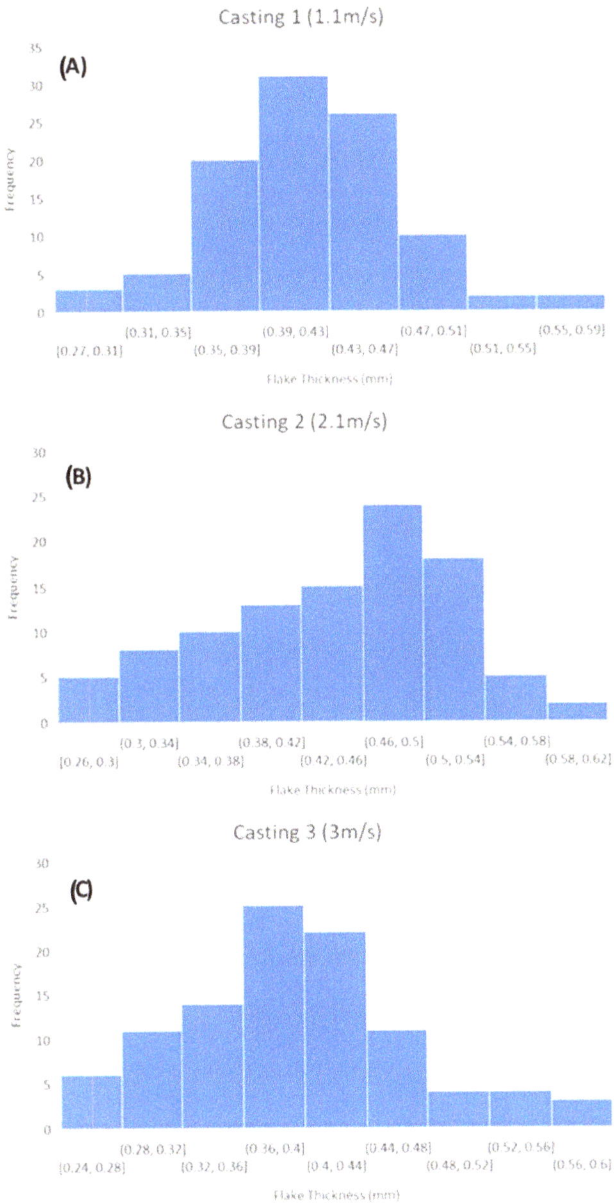


**Figure 3.** Flake thickness distributions from (**A**) Run 1, (**B**) Run 2, and (**C**) Run 3.

From Figure 4(A1), it can be seen that the thinnest flakes have a small nucleation region, a long region of columnar growth, and a small equiaxed region. The columnar grains are typically 1–2 µm in width, opposite the central nucleation zone. The maximum length of this zone is 235 µm, which reaches the free surface of the flake; however, the rest of the grains not directly in line with the nucleation zone become disoriented close to the free side of the flake. The median and thickest flakes (Figure 4(B1,C1), respectively), however, demonstrated very small regions of columnar growth (maximum length 140 µm and 179 μm, respectively) and predominantly consist of non-directional growth, with increased contrast in the image highlighting high levels of segregation observed in the median flake in particular. The phases that can be observed in these images are the 1:5H phase (white), the 1:7H phase (light grey), and the 2:17R/2:17H phases (dark grey).

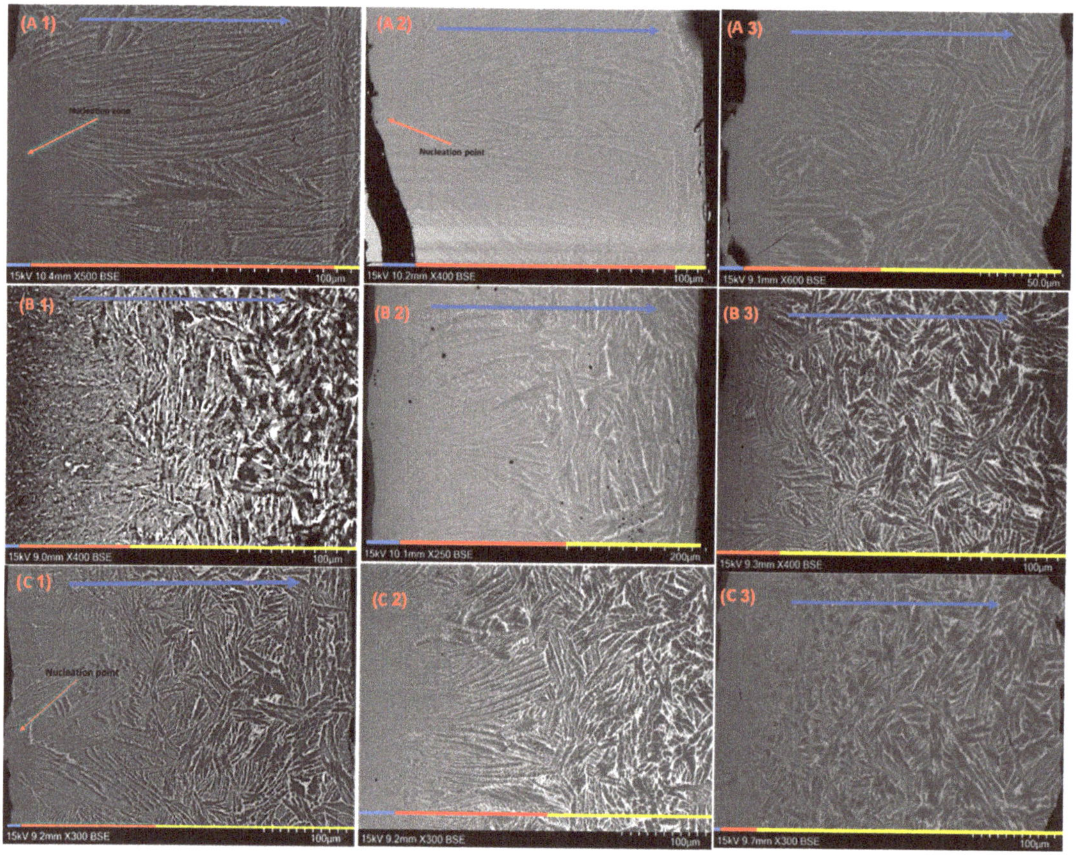


**Figure 4.** SEM micrographs of Run 1 from the (**A1**) thinnest, (**B1**) median, and (**C1**) thickest flakes; Run 2 from the (**A2**) thinnest, (**B2**) median, and (**C2**) thickest flakes; and Run 3 from the (**A3**) thinnest, (**B3**) median, and (**C3**) thickest flakes.

The microstructures of flakes from Run 2 show that the columnar growth regions in the median and thickest flakes (Figure 4(B2,C2), respectively) are longer than for Run 1 (~207 μm and 201 μm). This was expected as the higher wheel speed should lead to a higher cooling rate, promoting columnar growth. The flake taken from the thinnest section, however, did not show a flat surface on the free side, where a single nucleation point could be identified, and all columnar growth emanated from this point with a maximum length of 251 μm. This is likely to have occurred as a result of poor contact between the melt and the wheel during casting. It is highly likely that the molten alloy only contacted the wheel at this point on the first contact; it bounced/skipped and then gripped the wheel on the second contact. This means that initial solidification started on the first contact; the matrix phase nucleated and started growing, which was accelerated as the alloy made a second contact with the wheel. The non-directional region at the free side of the flake in this case is slightly larger (26–46 μm) than in the thinnest flakes from Run 1 (3–25 μm), due to the issues with contact with the wheel.

The microstructures of flakes from Run 3 are also shown in Figure 4. It can be seen that although the mean flake thickness was lower than in Runs 1 and 2 (see Figure 3), the microstructures are far from ideal. All three thicknesses show structures predominantly consisting of non-directional grains with very small regions of columnar growth. The thin flakes show a clear nucleation zone (Figure 4(A3)); however, it is less defined in the median and thickest flakes (Figure 4(B3,C3), respectively). This is likely due to poor adhesion/contact between the molten alloy and the wheel as it spins at high speed. The alloy is likely to have skipped or bounced off the wheel (as with the thinnest flake in Run 2); however, the contact in these cases was insufficient to maintain the required cooling rate for columnar growth, and the maximum length of columnar grains was just 112 μm.

The microstructures of anomalies taken from Runs 1 and 2 are shown in Figure 5a,b, respectively. These flakes demonstrate the issues caused by the turbulent flow of the melt onto the wheel during strip casting. Figure 5a shows a flake where the thickness is not uniform; towards the top of the image, the flake is almost twice the thickness that it is at the bottom. In this case, there is a very typical nucleation zone on the wheel side, a columnar growth region, and then non-directional growth. Towards the top of the image, the microstructure would be deemed to have approximately equal columnar growth and non-directional growth, whereas in the thinner section, the microstructure is predominantly columnar growth with minimal non-directional growth.

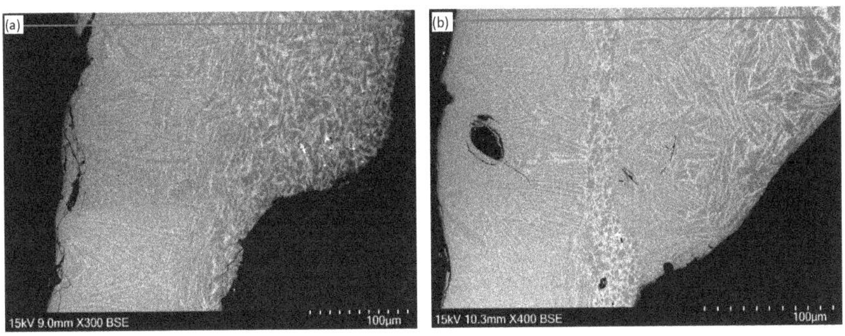


**Figure 5.** SEM micrographs of a non-uniform sample obtained from (**a**) Run 1 and (**b**) Run 2. The blue arrow indicates the cooling direction from the wheel side (**a**) to the free side (**b**).

In Figure 5b, the microstructure is much more irregular. It appears that a thin flake has initially been formed with a thickness of ~120 μm and a predominantly columnar structure. There is a large void within the flake, which was likely formed due to turbulence in the flow of the liquid onto the wheel. It appears from the micrograph that more liquid has flowed over the surface of the thin flake, which has created a band of highly segregated microstructure, likely due to re-melting of the top surface of the flake. The alloy that has flowed over the original flake has then solidified with some directional growth (left to right) and then non-directional growth towards the very right hand side of the image. This would suggest that as the secondary flow of melt solidified on a thin flake (~120 μm), the cooling rate was still sufficient to induce some level of columnar growth. It also gives further justification for the importance of controlling the flow of the alloy onto the wheel in order to obtain the desired microstructure.

Figure 6 shows XRD analysis of the as-received ingot as well as median flakes from each strip cast run, which was performed on coarse powders generated by grinding the flakes with a pestle and mortar. It can be seen that the traces of all strip cast runs are very different from the as-received ingot. The flakes from Run 1 show large fractions of 1:7H phase (denoted by *) with small amounts of 1:5H and minor fractions of 2:17R. This shows that the cooling rate was largely sufficient to form 1:7H directly during strip casting, although evidence of the remaining minority phases highlights that the casting parameters were not quite ideal as some heat treatment would be required to develop the

1:7H metastable phase throughout the entire structure. Run 2 is very similar; however, there is evidence of an additional peak at 48° 2θ, which corresponds to the presence of the 2:17H phase, which would be deleterious in further processing. Run 3 shows an increase in the phase fractions of 2:17R and 1:5H through increased peak intensity as well as the identification of new peaks for these phases.

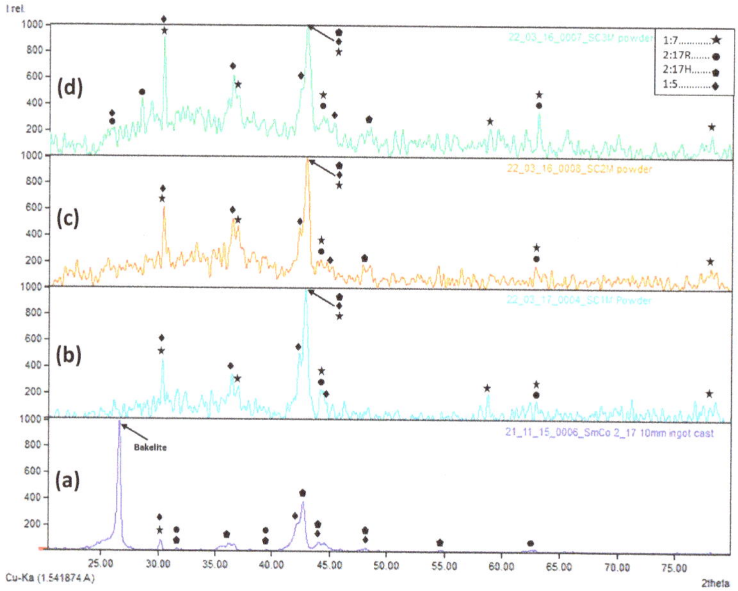


**Figure 6.** XRD traces for (**a**) as-received ingot and median flakes from (**b**) Run 1, (**c**) Run 2, and (**d**) Run 3.

XRD was also performed on the wheel side and free side of the complete flakes from the median batch from Run 1, as shown in Figure 7. These samples were also analysed using SEM and EDX to highlight the difference in microstructure on the flake surfaces and determine the compositional differences between the different phases, which are presented in Figure 8.

It can be seen that there is a significant difference in microstructure and phase balance between the as-received ingot, the wheel side of the flake, and the free side of the flake. This clearly indicates that there is a vast change in microstructure and phase balance as a result of strip casting. The change in microstructure through the cross section of the flake is also evident, as the peak locations and intensities vary from the wheel side to the free side of the flake. It was evident from the micrographs in Figure 4 that there were microstructural changes; however, the XRD analysis shows that this was not simply a change from columnar growth to equilibrium due to a change in cooling rate but also a change in phase balance as Sm and Cu rich phases are formed and segregate from the Fe and Co rich phases, which is evident from the BSE SEM images and EDX spectra in Figure 8. Interestingly, however, the XRD trace for the free side looks very different to that of the as-received ingot; this was partially due to a range of blue and gold coloured oxides being present on the surface as well as the fine, inhomogeneous microstructure. The peaks presented in Figure 7 do not accurately match those of the 1:5H, 1:7H, 2:17H, or 2:17R phases due to the surface oxides; hence, this data does not confirm the actual phase balance but clearly highlights a compositional difference between the two surfaces of the flakes. The EDX analysis shows that in general, the Sm and Cu content of the phases on the wheel side of the flake is higher than on the free side, and an inverse relationship is observed with Co and Fe.

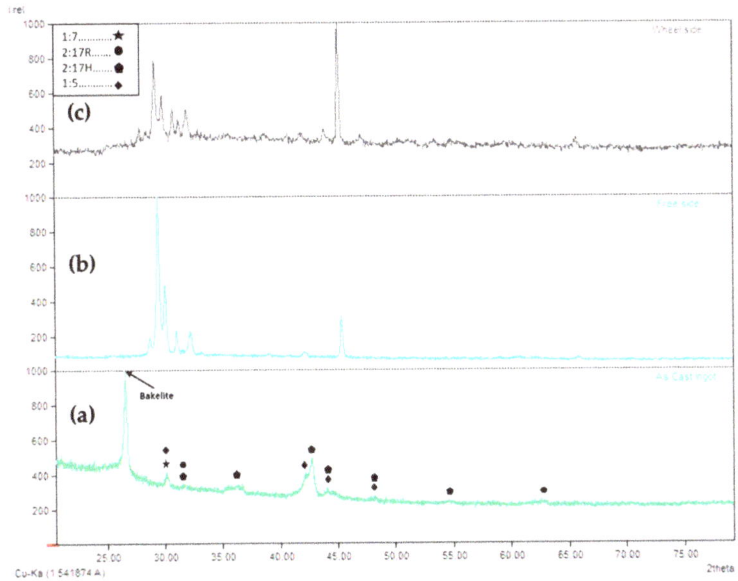


**Figure 7.** XRD traces for (**a**) the as-received ingot, (**b**) the free side, and (**c**) the wheel side of median flakes from Run 1.

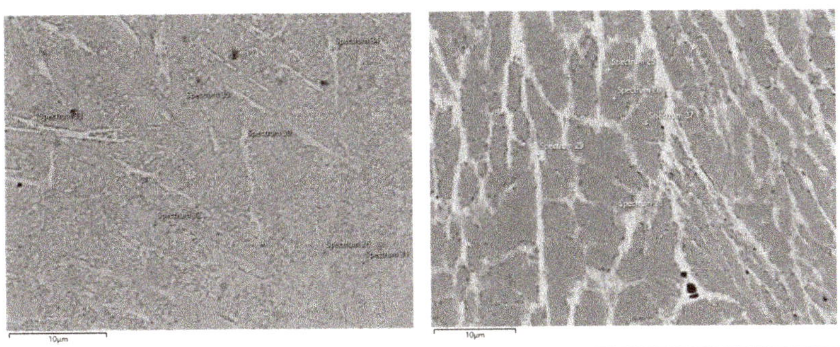

| | Sm (at%) | Co (at%) | Fe (at%) | Cu (at%) | Zr (at%) |
|---|---|---|---|---|---|
| **Wheel Side** | | | | | |
| Spectrum 30 | 15.71 | 53.84 | 18.00 | 7.34 | 5.12 |
| Spectrum 31 | 12.74 | 53.03 | 23.80 | 8.72 | 1.70 |
| Spectrum 32 | 12.21 | 53.71 | 24.85 | 6.28 | 2.48 |
| Spectrum 33 | 10.05 | 51.24 | 24.89 | 11.79 | 1.51 |
| Spectrum 35 | 12.92 | 52.42 | 21.74 | 10.08 | 2.85 |
| Spectrum 36 | 12.75 | 53.09 | 24.42 | 7.70 | 1.53 |
| **Free Side** | | | | | |
| Spectrum 25 | 7.02 | 54.37 | 23.76 | 13.01 | 1.36 |
| Spectrum 27 | 10.90 | 55.27 | 27.97 | 4.67 | 1.18 |
| Spectrum 28 | 11.69 | 54.80 | 26.87 | 4.80 | 1.63 |
| Spectrum 29 | 14.47 | 54.65 | 20.90 | 6.81 | 3.17 |

**Figure 8.** BSE SEM images of the wheel side (**left**) and free side (**right**) of the median flake from Run 1, along with EDX spectra from the points highlighted on the micrographs.

## 4. Discussion

The as-received cast alloys consist predominantly of the 2:17H phase with small amounts of the 2:17R, 1:5H, and 1:7H phases, and the microstructure demonstrated significant elemental segregation as a result of slow cooling in a water-cooled mould. The macrosegregation of Fe and Co to the dark phase and Cu and Sm to the light phase would require a homogenisation treatment to be removed before further processing into a magnet. For all strip cast alloys, however, the phase distribution changed substantially to consist predominantly of the metastable 1:7H phase with small amounts of 1:5H phase and trace 2:17R phase. Due to the substantially increased cooling rate, the microstructure typically consisted of a nucleation zone where the melt started to solidify in contact with the wheel and columnar growth along the thermal gradient from the wheel side of the flake towards the free side, similar to that demonstrated by Meng et al. [19] and Yang et al. [20]. However, in a number of flakes, there were regions where the cooling rate was not sufficient to maintain columnar growth, resulting in non-directional growth towards the free side of the flake and elemental segregation similar to that observed in the as-received cast alloy.

It was expected that as wheel speed increased, the flake thickness would be reduced as the molten alloy was pulled onto the wheel more quickly, which should result in thinner flakes consisting of columnar grains and the absence of non-directional growth. However, the faster wheel speed of 3.0 m/s produced a strip cast alloy with minimal columnar growth and predominantly non-directional growth and elemental segregation. Further analysis of Figure 4 showed that the contact surface between the wheel and the molten alloy was inconsistent as turbulence was introduced due to the high wheel speed. The sample that produced the most ideal microstructure was actually produced at the lowest wheel speed of 1.1 m/s in the thinnest flakes from the batch. The flake thickness was 252 μm, demonstrating a uniform microstructure consisting of a nucleation zone 16 μm thick, a columnar region of 222–235 μm, and non-directional growth of 3–25 μm. In terms of maximum columnar growth region, the thinnest flakes produced with a wheel speed of 2.1 m/s were 300 μm thick, with a 19 μm nucleation zone, 225–251 μm columnar region, and 24–46 μm non-directional growth. This would suggest that the maximum flake thickness should be 270 μm thick to include the nucleation zone and columnar region whilst avoiding non-directional growth where the cooling rate cannot be sustained. Considering the microstructure across all samples suggests that control of the melt flowing onto the wheel and the contact surface between the melt and the wheel is more important for obtaining the desired microstructure than the rotational speed of the wheel. This was further highlighted by the microstructural anomalies presented in Figure 5, where turbulent flow led to inconsistent flake thickness and dual-microstructures within a flake. In a commercial system where >300 kg of alloy is produced in a single casting, small amounts of alloy with undesirable microstructures are unlikely to be detected and will be mixed in with the optimised material. However, in pilot-scale strip casters, there is more likely to be differential contact and cooling within a batch due to the low amount of alloy in the system (<5 kg) and sampling of the flakes for analysis after casting, resulting in a higher proportion of undesirable alloy. This has not been identified in most literature studies, as they utilise melt spinners that use pressurised gas to consistently eject the material onto the wheel rather than gravity feed through a series of tundishes. This is therefore not representative of commercial strip casting systems.

The XRD analysis presented in Figure 6 shows that the cooling rate was largely sufficient to form 1:7H directly during strip casting, although evidence of the remaining minority phases highlights that the casting parameters were not quite ideal as some heat treatment would be required to develop the 1:7H metastable phase throughout the entire structure. Figure 7 showed a difference in key peak intensities between the wheel side of the flake and the free side, which confirmed that the change in microstructure from the nucleation zone to the columnar region to non-directional growth is accompanied by a significant change in phase balance between the desired and detrimental phases. This was confirmed by the SEM and EDX analyses presented in Figure 8. Overall, this would suggest that further optimisation of the tundish design to better control alloy flow onto the wheel would likely result in the desired columnar microstructure and phase balance of 1:7H and 1:5H without non-directional growth or unwanted minority phases. Alternatively, increasing the surface roughness of the wheel through shot blasting may increase adhesion between the wheel and the melt and result in more consistent cooling throughout the alloy.

When comparing these findings to those of Liu et al. [16] and Liu et al. [18], the flakes in this work show lower homogeneity in terms of phase balance and microstructure. Liu found that their strip cast ribbons contained the majority 1:7H phase with small traces of 1:5H, whereas this work found the majority 1:7H phase with fractions of 1:5H, 2:17H, and 2:17R. As Liu used the melt spinning technique to replicate strip casting, it is very likely that uniform contact was observed between the melt and the wheel as a result of the force applied by pressurised gas. This means that the cooling rate will be much more uniform than in gravity fed strip casting, as presented in this paper. However, homogeneity in the cooling rate means that the cross-sectional microstructure presented non-directional growth rather than columnar growth [16,18]. The fine-grained strip cast ribbons also led to lower remanence in fully processed magnets as it was not possible to produce single crystal particles by milling, which may not be a problem with the strip cast alloys presented in this work. Meng et al. [19] demonstrated strip cast flakes that were 575 μm thick, where fine grains dominated 445 μm of the thickness and columnar grains the remaining 130 μm. This also led to a reduction in the magnetic properties of processed magnets compared to ingot starting materials, which was attributed to the high proportion of fine-grained material leading to polycrystalline particles after milling. As the strip casting parameters were not given and limited microstructural analysis was presented, no direct comparisons can be made regarding the optimisation of the strip casting process. Yang et al. and Yang et al. [20,21] demonstrated very similar findings to Meng et al. [19], showing that slower wheel speeds in strip casting led to better magnetic properties; however, as Meng et al., were utilising the full manufacturing route for magnet processing and were not aiming to circumvent the long heat treatment as would be the aim of this work. Likewise, direct comparison with Zheng et al. [22] is limited as they did not optimise the strip casting process; however, they obtained a similar combination of phases of the alloy but were not able to demonstrate directional columnar growth from the wheel side to the free side of the flake as presented in this work.

## 5. Conclusions

The work presented in this paper has shown that it is possible to produce Sm2TM17-type alloys directly from strip casting to give a microstructure predominantly consisting of the $SmTM_7$ hexagonal phase. Lower wheel speeds of 1.1 m/s and 2.1 m/s yield alloys with a higher proportion of the 1:7H phase with a more homogenous structure, which was attributed to better adhesion between the molten alloy and the water-cooled wheel during solidification. A wheel speed of 3 m/s led to significant elemental segregation within the microstructure due to poor adhesion of the melt to the wheel during solidification. The highest length columnar growth region was found to be 251 μm, consisting mainly of the 1:7H phase, which could be a maximum flake thickness to achieve a fully columnar structure. It was shown that controlling the contact between the melt and the wheel is more important than wheel speed alone to ensure that contact and cooling is sufficient throughout. Further optimisation of the process, for example, by changing tundish design and gate height to better control flow of the alloy onto the wheel or shot blasting the wheel to increase texture and hence aid gripping of the alloy to the wheel, could lead to fully homogenous 1:7H structures, removing the need for the homogenisation and solutionising stages of magnet production.

**Author Contributions:** Conceptualization, R.S.; Methodology, R.S., J.G.-F., A.A. and M.B.; Validation, A.A., J.G.-F. and M.B.; Formal Analysis, R.S. and J.G.-F.; Investigation, R.S., A.A., J.G.-F. and M.B.; Resources, R.S.; Writing—Original Draft Preparation, R.S. and J.G.-F.; Writing—Review and Editing, R.S., A.A., J.G.-F. and M.B.; Visualization, A.A. and J.G.-F.; Supervision, R.S.; Project Administration, R.S.; Funding Acquisition, R.S. All authors have read and agreed to the published version of the manuscript.

**Funding:** This research received no external funding.

**Data Availability Statement:** The raw data supporting the conclusions of this article will be made available by the authors on request.

**Acknowledgments:** The authors would like to acknowledge Less Common Metals Ltd. for supply of the master ingots used in this study.

**Conflicts of Interest:** The authors declare no conflicts of interest.

## References

1. Yi, J.-H. Development of samarium–cobalt rare earth permanent magnetic materials. *Rare Met.* **2014**, *33*, 633–640. [CrossRef]
2. Dormidontov, N.A.; Kolchugina, N.B.; Profof'ev, P.A.; Zheleznyi, M.V.; Milov, Y.V.; Dormidontov, A.G.; Bakulina, A.S. Structural constituents and phases in high-coercivity (Sm,Zr)(Co,Cu,Fe)z alloys for permanent magnets. *Russ. Metall.* **2022**, *2022*, 505–511. [CrossRef]
3. Hadjipanayis, G.C.; Tang, W.; Zhang, Y.; Chui, S.T.; Liu, J.F.; Chen, C. High temperature 2:17 magnets: Relationship of magnetic properties to microstructure and processing. *IEEE Trans. Magn.* **2000**, *36*, 3382–3387. [CrossRef]
4. Shang, Z.F.; Zhang, D.T.; Altounian, Z.; Xie, Z.H.; Qiao, P.B.; Liu, W.Q. Effect of ingot cooling rate on Cu distribution and magnetic properties of Sm(Cobal $Fe_{0.28}Cu_{0.07}Zr_{0.03})_{7.6}$ magnets. *AIP Adv.* **2019**, *9*, 125–142. [CrossRef]
5. Shang, Z.; Zhang, D.; Xie, Z.; Wang, Y.; Haseeb, M.; Qiao, P. Effects of copper and zirconium contents on microstructure and magnetic properties of Sm(Co,Fe,Cu,Zr)z magnets with high iron content. *J. Rare Earths* **2021**, *39*, 160–166. [CrossRef]
6. Xue, Z.Q.; Liu, L.; Liu, Z.; Li, M.; Lee, D.; Chen, R.J. Mechanism of phase transformation in 2:17 type SmCo magnets investigated by phase stabilization. *Scr. Mater.* **2016**, *113*, 226–230. [CrossRef]
7. Yan, A.; Gutfleisch, O.; Handstein, A. Microstructure, microchemistry, and magnetic properties of melt-spun magnets. *J. Appl. Phys.* **2003**, *93*, 7975. [CrossRef]
8. Horiuchi, Y.; Hagiwara, M.; Okamoto, K.; Kobayashi, T.; Endo, M.; Kobayashi, T.; Nakamura, T.; Sakurada, N. Effects of solution treated temperature on the structural and magnetic properties of iron-rich $Sm(Co,Fe,CuZr)_z$ sintered magnet. *IEEE Trans. Magn.* **2013**, *49*, 3221–4224. [CrossRef]
9. Musa, M.; Zhou, X.; Song, X.; Yuan, T.; Jia, W.; Wang, J.; Ma, T. Fe content influence on the microstructure of solution-treated Sm-Co-Fe-Cu-Zr alloys. *Intermetallics* **2021**, *129*, 107049. [CrossRef]
10. Hu, M.; Kang, D.; Zhang, T.; Liu, B.; Xi, L.; Cao, J.; Xu, C.; Zuo, S.; He, Y.; Jiang, C. Single 1:7H solid-solution phase achieving in iron-rich Sm-Co-Fe-Cu-Zr magnets. *J. Alloys Compd.* **2023**, *945*, 169373. [CrossRef]
11. Cui, C.; Liu, X.; Nlebedim, C.I.; Cui, J. Mechanically strengthened heterogenous Sm-Co sintered magnets. *J. Alloys Compd.* **2022**, *904*, 163937. [CrossRef]
12. Campbell, A. Investigations into the Optimisation of the Commercial Processing Route for $Sm_2Co_{17}$ Type Permanent Magnets. Ph.D. Thesis, University of Birmingham, Birmingham, UK, 2018.
13. Appleby, A. Investigating the Viability of Strip Casting Samarium Cobalt 2:17. Master's Thesis, University of Birmingham, Birmingham, UK, 2021.
14. Zapuskalov, N. Comparison of Continuous Strip Casting with Conventional Technology. *ISIJ Int.* **2003**, *43*, 1115–1127. [CrossRef]
15. Pei, W.; He, C.; Lian, F.; Zhou, G.; Yang, H. Structures and magnetic properties of sintered Nd–Fe–B magnets produced by strip casting technique. *J. Magn. Magn. Mater.* **2002**, *239*, 475–478. [CrossRef]
16. Liu, Z.; Liu, L.; Chen, R.J.; Sun, Y.L.; Lee, D.; Yan, A.R. Optimization of temperature coefficient of remanence and magnetic properties of sintered $Sm_{0.7}Dy_{0.1}Gd_{0.2}(CobalFe_{0.2}Cu_{0.08}Zr_{0.025})_{7.2}$ magnets prepared by strip-casting technique. *IEEE Tran. Magn.* **2013**, *49*, 5599–5602. [CrossRef]
17. Meakin, J.P. Development of a pilot scale strip casting system to produce optimised alloys for neodymium-iron-boron magnets. Ph.D. Thesis, University of Birmingham, Birmingham, UK, 2015.
18. Liu, L.; Liu, Z.; Chen, R.; Liu, X.; Yan, A.; Lee, D.; Li, W. Effect of strip casting on microstructure and magnetic properties of 2:17 type Sm-Co sintered magnets. *IEEE Trans. Magn.* **2014**, *50*, 2101704. [CrossRef]
19. Meng, C.Z.; Zhang, D.T.; Jia, J.H.; Yang, J.J.; Teng, Y.; Zhang, H.G.; Li, Y.Q.; Liu, W.Q.; Yue, M. Effect of casting methods on magnetic properties of Sm(CoFeCuZr)z magnet. In Proceedings of the 67th Annual Conference on Magnetism and Magnetic Materials, Minneapolis, MN, USA, 31 October–4 November 2023; Volume 13.

20. Yang, Y.; Zhang, D.; Zhang, H.; Li, Y.; Meng, C.; Teng, Y.; Liu, W.; Yue, M. Combination strategy for high-performance $Sm(CoFeCuZr)_z$ sintered permanent magnet: Synergistic improvement of the preparation process. *Acta Mater.* **2023**, *251*, 118901. [CrossRef]
21. Yang, J.; Zhang, D.; Zhang, H.; Shang, Z.; Wang, H.; Meng, C.; Liu, W.; Yue, M. Effect of hydrogen pressure on hydrogenation and pulverization behaviour of $Sm(CoFeCuZr)_z$ ingot and strip casting flake. *J. Alloys Compd.* **2023**, *930*, 167427. [CrossRef]
22. Zheng, M.; Bi, Y.; Hou, Z.; Wang, C.; Wang, J.; Wang, L.; Duan, Z.; Zhang, B.; Fang, Y.; Zhu, M.; et al. Enhanced hydrogen adsorption capability of 2: 17-type Sm-Co strips via microstructure modification. *J. Alloys Compd.* **2024**, *987*, 174182. [CrossRef]

metals

MDPI

Article

# A Methodology to Define the Niyama Criterion Reinforced with the Solid Fraction Analysis: Application to Sand Casting of Steel Bars

**María Carmen Manjabacas [1,2] and Valentín Miguel [1,2,*]**

1 Higher Technical School of Industrial Engineering of Albacete, University of Castilla-La Mancha, 02071 Albacete, Spain; mcarmen.manjabacas@uclm.es

2 Regional Development Institute, Science and Engineering of Materials, University of Castilla-La Mancha, 02071 Albacete, Spain

* Correspondence: valentin.miguel@uclm.es

**Abstract:** Niyama and solid fraction criteria are used to predict the solidification porosity and microporosity in computing simulation of casting processes. The solid fraction permits us to determine the areas that solidify last and that are a candidate for presenting porosity if a feeding system is not correctly designed. The Niyama criterion is locally obtained based on the thermal and cooling gradients at a point of the liquid casting. The Niyama value at a casting point varies rapidly from low rates to high ones during the last part of the metal solidification, which demands that the percentage of solidification of the metal is defined to determine the Niyama number. In addition, the Niyama threshold that establishes the soundness of the workpiece can vary according to the nature of the metal or the casting system. In this paper, a methodology to determine the solidification percentage is presented. The method is based on the Niyama number evolution during the solidification process at different key points. These points are validated by the solid fraction criterion as healthy or, on the contrary, as candidates for containing porosity. In addition, some considerations of the solid fraction criterion are visited since the threshold value for which the isolation of the last solidification areas can be defined is not clear. The research is validated by the empirical casting criteria existing in the literature for obtaining sound parts and applied to low-carbon steel bars produced by sand casting.

**Keywords:** sand casting; Niyama criterion; solid fraction; FEM casting simulation

## 1. Introduction

Porosity existing in cast parts can be harmful depending on the size, number and distribution of the pores existing in them [1–3]. Only in the cases in which pores are small and very uniformly distributed may their influence on the mechanical properties of the part be not significant, for example, in the casting of thin plates [4]. There are several origins of the pore formation: gases coming from the chemical reaction of different elements in the casting system, air introduced with the metal pouring, entrapment of air existing in the mold cavity, gases dissolved in the metal at high temperatures and, finally, empty areas created as the metal shrinks during the solidification. There are different solutions to minimize pore formation depending on its origin, and, in any case, it is important to predict it.

Shrinkage porosity appears in parts of the casting that are isolated from the rest as it is impossible to compensate for the solidification contraction by any medium. By FEM simulation, controlling solid fraction evolution of the metal during its solidification permits us to detect the parts of the metal casting that solidify last. Thus, according to the system, an analysis of critical areas can be stated and modifications to the design formulated [5]. Other criteria based on thermal parameters can be implemented by FEM, such as the local temperature gradient, G, the metal cooling rate, R, the solidification velocity, vs, the total freezing time, tf, and the local solidification time, ts. Different criteria based on a combination of those parameters have been defined and experimented with in the literature [6,7]. With these criteria, it is possible to obtain shrinkage porosity indicators at each point of the casting during the solidification process by FEM software. The most known of these criteria is the Niyama number, Ny, defined as $Ny = G/\sqrt{R}$ [8]. Niyama et al. demonstrated that points of the casting with an Ny value lower than that which is critical should present microporosity. Even above the critical number, a range of values reveals the existence of microporosity [9]. The criterion was built under the statement that links the Ny value and the pressure drop in the mushy zone [8,10]. Although the Ny critical value depends on the material, and, particularly, Niyama et al. obtained values in the range 0.9–1.5 $K^{1/2}\ min^{1/2}\ cm^{-1}$, they stated for steel a value of 1 $K^{1/2}\ min^{1/2}\ cm^{-1}$ as the critical one.

The critical Niyama value depends on the units in which it is defined, and sometimes Ny expressed according to its classical definition must be multiplied by 775 or divided by 1.29 if expressed as $K^{1/2}\ s^{1/2}\ m^{-1}$ or $K^{1/2}\ min^{1/2}\ mm^{-1}$, respectively. The Niyama criterion has some constraints: it does not consider pore nucleation and gravity action on it; the Ny threshold value is shape dependent [11] and generally unknown; and it is a qualitative criterion that does not provide the quantity of shrinkage porosity that forms [6,12]. Due to the last limitation, Carlson and Beckermann [12] defined a dimensionless Niyama criterion. Other authors proposed other parameters, such as the rate of transition of the alloy from a semisolid to solid state, establishing for cast iron a strong relationship between that parameter and the formation of shrinkage cavities [13].

Definitely, each material will have its own critical Ny value, usually determined by specific casting experiments [14]. In addition, Ny value strongly depends on the temperature at which it is evaluated, and, thus, it is important to define that. Some authors recommend evaluating Ny at a temperature that is 10% of the solidification range, $T_L - T_S$, above the solidus temperature, i.e., $T_{Ny}^{0.9} = T_S + 0.1(T_L - T_S)$ [15], according to the criterion defined by Carlson and Beckermann [16,17]. Nevertheless, based on some experiences, other authors state that the temperature at which Ny should be obtained could be different, although near the end of the metal solidification [14]. Finally, although some advances have been made in determining the threshold Ny for different alloys, this value is not known for all alloys or all steel [18].

Some FEM simulation software gives an Ny value for each point considering only one temperature, which is sometimes unknown or difficult to select with respect to the solidification percentage that it represents [19]. Other applications give Ny for any temperature, for example, VULCAN (Quantech ATZ S.A., Barcelona, Spain). Inspire Cast (Altair Engineering Inc., Troy, MI, USA) permits the Niyama value to be evaluated once the registers of the thermal gradient and the cooling rate have been obtained since only a unique value for $T_{Ny}^{0.85}$ is given directly through the results interface. This is a valuable option since it is possible to analyze the Ny value evolution during the solidification of any selected point. Moreover, this permits $T_{Ny}$ to be fixed according to other complementary analyses. Wang et al. [20] employed FEM software for casting simulation based on a CAD/CAE integrated system to optimize the casting process by the interaction among the different existent moduli. That system could calculate and display hot spots to optimize the risering system by means of an active intelligent mode. Definitely, the advances of the casting numerical simulation permit the casting production to be guided [21], and specific procedures have been applied by some authors in order to eliminate or reduce the shrinkage porosity [22].

In this work, a sand casting process for obtaining a low-carbon steel bar is analyzed.

The empiric fundamentals taken into consideration to calculate the gating system are established in [5]. The filling system was dimensioned to fulfill the requirement of achieving a solid fraction that equals zero at the end of the mold filling. The casting system was analyzed by FEM simulation, and it was probed so that the filling system led us to obtain a part without typical defects linked to the filling process, i.e., the empirical and computer simulation results were coherent [5]. The feeding system was based on the typical thermal and volume criteria and shape factor design established in the literature and based on empirical experiments to obtain sound pieces [4,23–28]. The considerations applied are established in Appendix A.

The length of the bar permitted the location of several risers to be analyzed. With respect to the feeding system, the dimension, number and location of the risers were experimented on by FEM, taking into account the solid fraction criterion. Aspects such as the mold end effect and the real heating transmitter surfaces were taken into account. For these, different solutions were analyzed, and the solution corresponding to the empirical one was probed as the optimal one [5].

The action of the filling system as a functional riser was demonstrated to be negligible since the gate solidified very fast, and the rest of the filling system remained isolated during the early solidification. In addition, it was demonstrated that the riser part joints work in a correct way, avoiding the isolation of the molten metal in the riser from the part during the solidification process; however, the riser necking sections are probably not enough to provide the necessary material into the part.

A deep analysis related to the application of the Niyama criterion was carried out. For that, the coherence between the solid fraction and Niyama criterion was taken into account to define the solidification considerations at different points of the casting system. The evolution of the Niyama number during the solidification was analyzed at those points, and, according to the expected results, the analysis criteria were defined. In addition, a methodology was planned to define the thresholds of the Niyama and solid fraction criteria for a correct analysis.

## 2. Materials and Methods

A sand casting system for producing a 1.0331 low-carbon steel bar, 1300 mm long, was considered for simulation by FEM by VULCAN® (Quantech ATZ S.A., Barcelona, Spain) and Inspire Cast (Altair Engineering Inc., Troy, MI, USA). The section of the bar was $70 \times 70$ mm$^2$. A green sand mold was selected; the filling and the feeding systems were designed according to classical rules, leading to sound pieces being obtained. The details of the system can be observed in Figure 1. The mold has a thermal conductivity that is variable with the temperature, according to Table 1. The conductive and convective heat transfer coefficients of the mold are also included in Table 1 [29]. The chemical composition of the steel is defined in Table 2, and its thermal properties are collected in Table 3 [29]. The solid fraction, SF, for the steel varies linearly from 1772.7 K, SF = 1, to 1800.2 K, SF = 0. The dynamic viscosity also varies linearly from 0.0059 to 0.0067 kg/(m·s) and between 1873 and 1825 K, respectively [29].

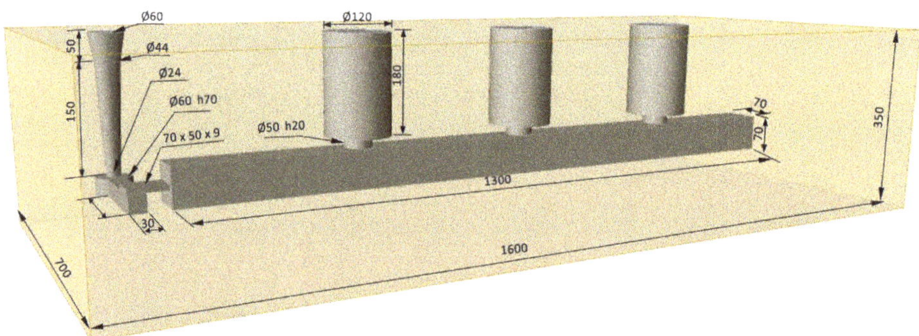


**Figure 1.** Sand casting system to obtain a low-carbon steel bar.

With respect to the casting technological conditions, it can be highlighted that the metal was overheated by 100 K, i.e., the pouring temperature was 1900 K. The initial temperature of the mold was 293 K, i.e., room temperature. The filling time was 15 s, which implies consideration of the threshold of velocities recommended by some authors to avoid turbulences during the filling process [4]. From this point of view, the well, i.e., the sprue–runner transition, designed in the system is not the best, but it was selected for its simplicity since no influence was expected on the goal of the present research.

**Table 1.** Thermal properties of the green sand mold considered in FEM simulation (from Materials library of Inspire Cast) [29].

| Thermal Conductivity W/(m K) | Conductive Heat Transfer Coefficient W/($m^2$ K) | Convective Heat Transfer Coefficient W/($m^2$ K) |
|---|---|---|
| 0.5 from 298.15 to 473.15 K<br>0.51 at 673.15 K<br>0.59 at 973.15 K<br>0.77 at 1273.2 K | 400 | 1000 |

**Table 2.** Chemical composition (% weight) of 1.0331 carbon steel.

| C | Si | Mn | P | S |
|---|---|---|---|---|
| ≤0.10 | Traces | 0.30–0.60 | 0.045 | 0.045 |

**Table 3.** Thermal conductivity of 1.0331 carbon steel (Materials library of Inspire Cast) [29].

| Thermal Conductivity (W/(mK)) |
|---|
| 27.6 at 1123 K |
| 27.2 at 1223 K |
| 28.85 at 1323 K |
| 29.85 at 1423 K |
| 29.7 at 1523 K |
| 29.85 at 1723 K |
| 30 at 1873 K |

*FEM Simulation of Solid Fraction during the Solidification*

FEM simulations were carried out by means of Inspire Cast (Altair Engineering Inc., Troy, MI, USA). This software permitted us to obtain, among other results, the solid fraction evolution, SF, during the solidification of the molten metal into the mold cavity. To analyze the solid fraction evolution, it is important to fix a threshold above which the liquid material existing in the mushy areas cannot go through the interdendritic channels from one part to another. The level of solidification at which this happens depends on the metal's natural property, and it is very important to determine the reference value since it can influence the results obtained. In Figure 2, the SF evolution during the solidification can be observed for two different threshold levels, 0.70 and 0.90. For the analysis, it should be considered that the white color represents the areas in which the SF was higher than 0.70 or 0.90. Usually, most metals are probed to have an SF threshold value between these two. As can be observed, there was hardly any influence of the level considered in the analysis on the critical isolated points of the molten metal that remain isolated during the last instants of the solidification process (areas rounded in red color in Figure 2). The only difference found was the size of the critical areas, which, as expected, was slightly greater for the lowest threshold experienced, i.e., 0.70. Logically, the reference solidification times in the analysis were lower for the minor SF limit. Only if the SF limit value was very low, could a substantial qualitative difference be observed in the SF analysis, shown in Figure 3. In Figure 3, for an SF reference of 0.30, it can be observed that some areas below the risers turn into isolated zones and might constitute critical points that present metal empty areas as a consequence of the shrinkage phenomenon that would not be compensated by molten metal in the neighborhood. Nevertheless, this last analysis lacks interest since a critical SF of 0.30 is excessively low.

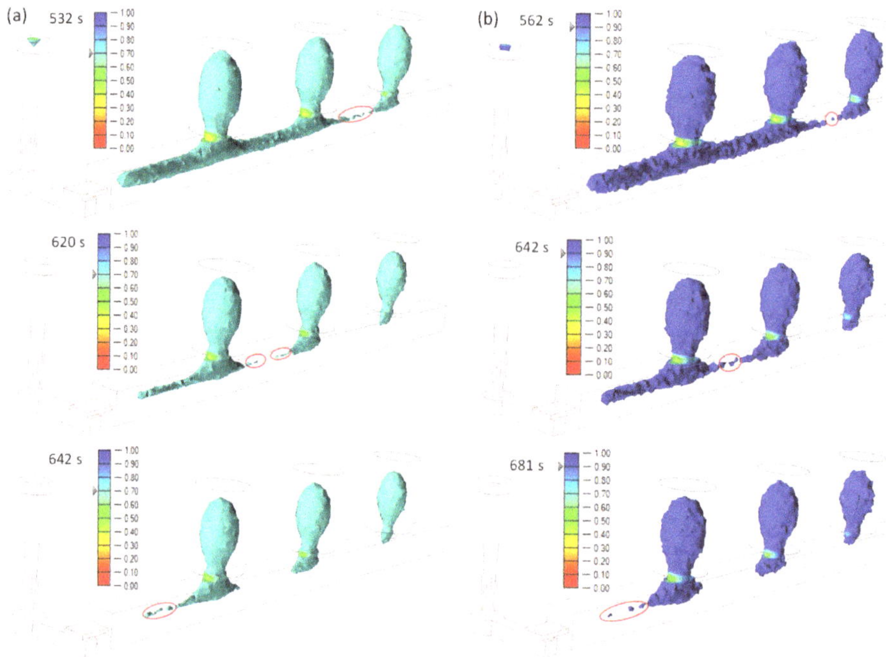


**Figure 2.** Solid fraction evolution considering two reference values that establish the SF limit above which no liquid transport is allowed: (**a**) SF reference value equals 0.70 and (**b**) SF reference value equals 0.90.

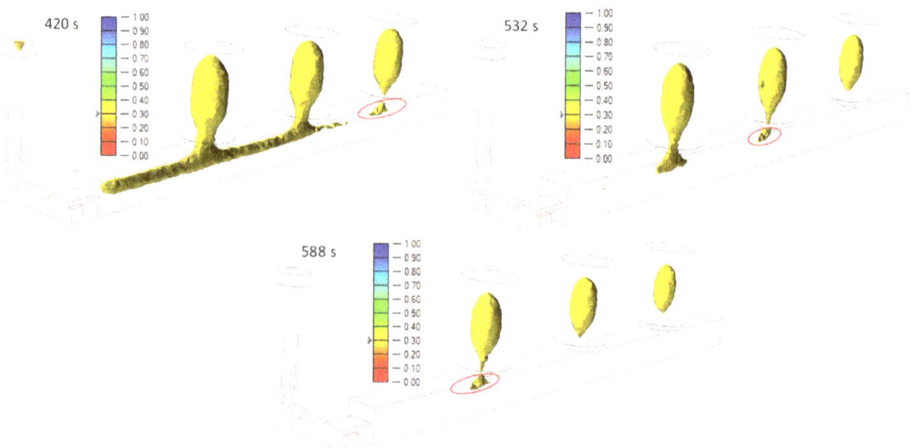


**Figure 3.** Three different instants of the solid fraction evolution analysis taking into consideration an SF reference value of 0.30. The red laces indicate isolated regions with an SF lower than 0.30 during the solidification.

### 3. Description of the Methodology Employed for Ny Analysis

Seven points of the casting system were selected for the analysis, shown in Figure 4i. All the points correspond to the central section of the casting system since the material belonging to this section solidifies last. Points 1 and 2 are located in places that are supposed to be covered by the feeding distance according to the empirical criteria applied; the only difference is that point 1 remains just in the center of the distance inter-risers, and point 2 is located at the extreme of the bar just out of the end workpiece zone. Point 3 is intermediate to the gate and the first riser, and nearer to the filling system, just where the SF criterion establishes the possibility of finding areas that have experienced shrinkage. The location of the first riser is at a distance from the extreme greater than the empirical rules recommend. Point 4 is placed just at the top of a riser in which, according to the SF criterion, the solidification takes place before that in the center of the riser. This means no shrinkage porosity is expected in this area. Point 5 is located just under a riser, and it is expected to be sound if the neck riser is correctly designed. Point 6 is just into the last solidification area in the riser that will remain totally isolated at the end of the solidification process. This supposes that a large empty area will form there, and the point selected should present a low Ny value. Finally, point 7 is into the end workpiece effect of the bar, and, according to the directional solidification fundamentals, it is expected to be a sound point.

Thus, the methodology consists of obtaining the Ny value at different $T_{Ny}$ values, defined according to Equation (1).

$$T_{Ny}^{1-\delta} = T_S + \delta(T_L - T_S) \tag{1}$$

The values adopted for δ were 0.2, 0.15, 0.10 and 0.05, which means that Ny was determined at temperatures at which the solidification process was completed up to 80, 85, 90 and 95%, approximately. Taking into consideration the expected results supported by the SF analysis, shown in Figure 2, it will be decided whether the Niyama values obtained show some coherence.

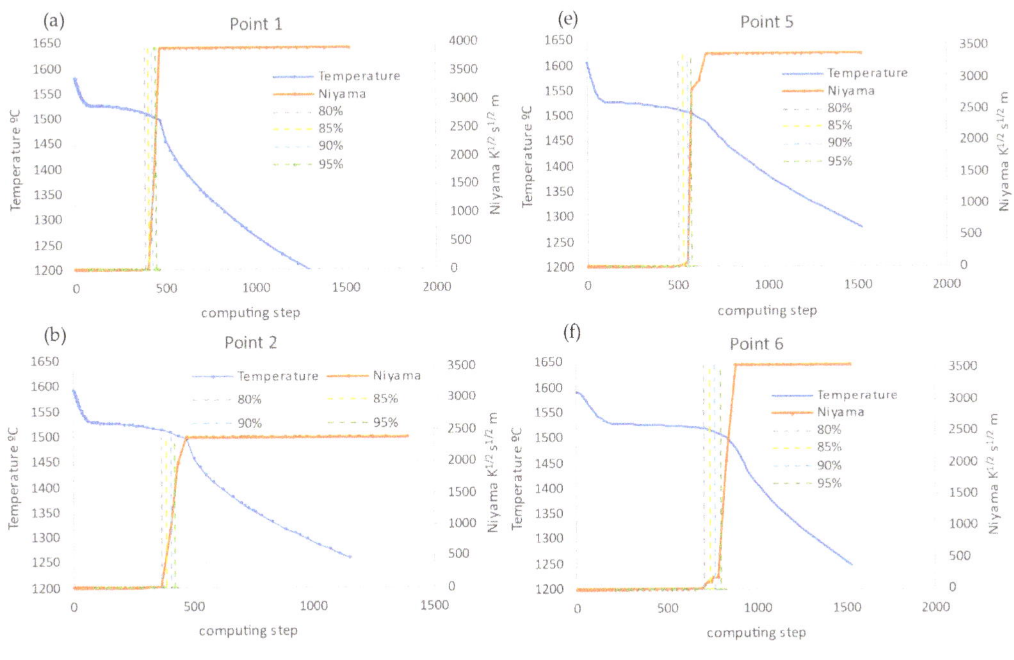


**Figure 4.** *Cont.*

(c)

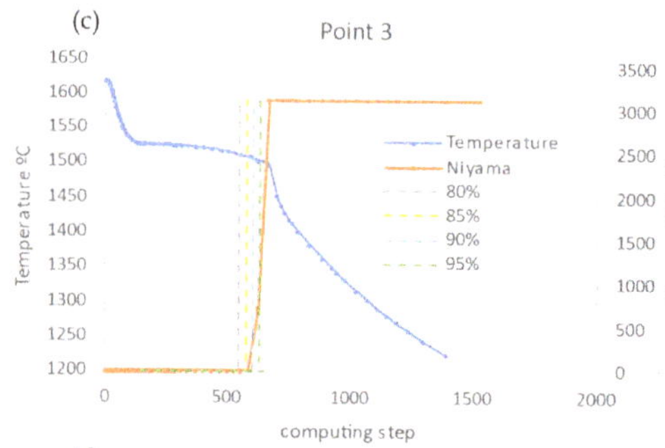


(g)

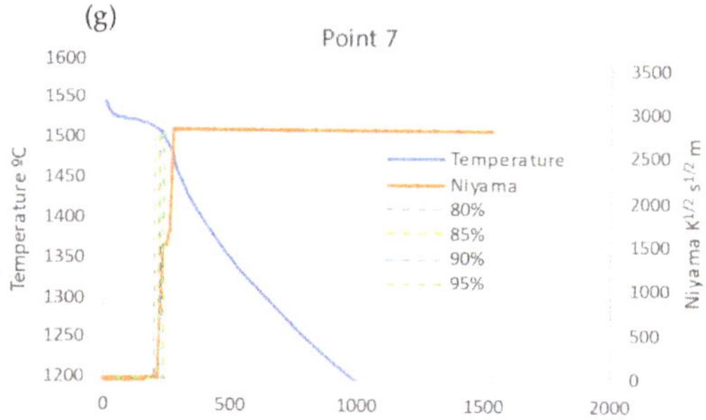


(d)

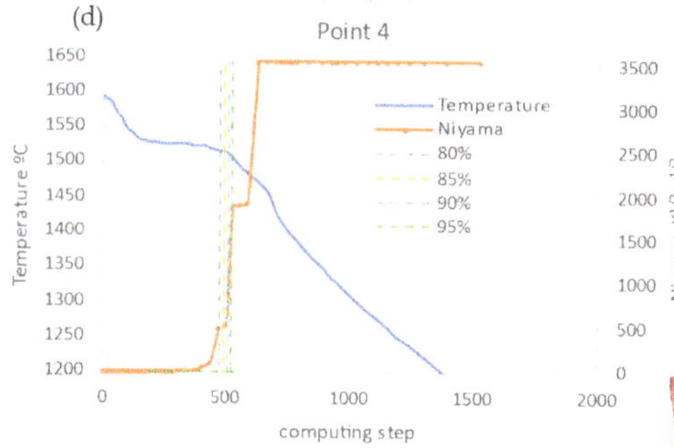


(h)

| | Niyama ($K^{1/2} s^{1/2} m^{-1}$) at different $T_{Ny}$ (%) | | | |
|---|---|---|---|---|
| | 80 | 85 | 90 | 95 |
| P1 | 8.4 | 168.3 | 1087.5 | 2125.7 |
| P2 | 98 | 532.9 | 984 | 1541 |
| P3 | 0 | 3.93 | 294.5 | 868 |
| P4 | 491.8 | 506.3 | 818.5 | 1773 |
| P5 | 9.5 | 31.4 | 307.7 | 2548.4 |
| P6 | 52.1 | 111.5 | 180.8 | 817.3 |
| P7 | 12.68 | 106.27 | 949.2 | 1481 |

(i)

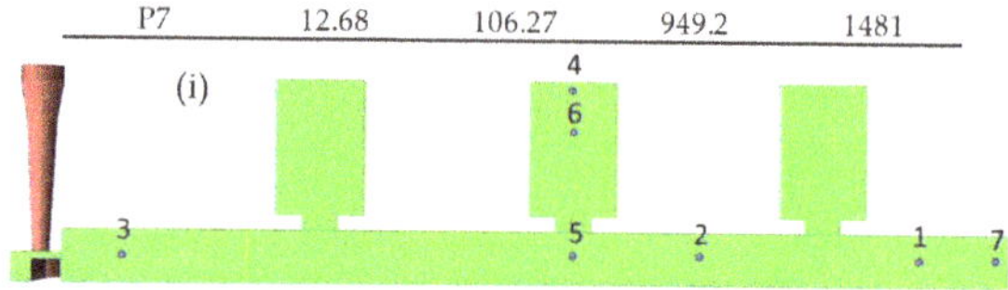


**Figure 4.** Metal cooling temperature curves and Niyama value evolution for the different points 1 to 7 (**a–g**) of the casting system (**h**); (**i**) indicates the Ny values corresponding to every point at the temperatures of $T_{Ny}^{0.8}$, $T_{Ny}^{0.85}$, $T_{Ny}^{0.9}$ and $T_{Ny}^{0.95}$.

## 4. Results and Discussion

In Figure 4, the metal cooling temperature curves belonging to the points analyzed are drawn in a blue color. The starting temperature corresponds to the instant at which the filling process has just finished, and, logically, it is not the same for the different points considered. As expected, point 3 presents the highest temperature after the filling process since the hottest metal enters through the first extreme of the bar, near point 3. The values corresponding to the other points are explained similarly, with 7 and 1 being the points with the lowest initial temperature just after the mold filling. Thus, the maximum difference in temperature between points 3 and 7 is 75 °C. The Ny evolution during the solidification is represented in an orange color. It can be seen that an abrupt change from 0 to very high values exists in the last part of the metal solidification, i.e., at temperatures near $T_s$, or the solvus temperature. Thus, it is very difficult to set graphically the Ny value corresponding to a determined temperature in the last part of the metal solidification process. In detail, it is necessary to determine the time step corresponding to the temperature $T_{Ny}^{1-\delta}$ in the metal cooling curve and to obtain the value of the Ny for this time step in the Niyama curve. In Figure 4, vertical dotted lines intersecting the *x*-axis represent the values corresponding to the time step of the metal cooling temperature function at the *y*-axis values of $T_{Ny}^{0.8}$, $T_{Ny}^{0.85}$, $T_{Ny}^{0.9}$ and $T_{Ny}^{0.95}$. Thus, the intersections of the dotted lines with the Niyama function permit Ny values to be found for the temperatures indicated before. Due to the difficulty of determining graphically the Ny values corresponding to each point i at temperatures near the end of solidification, Nyi, those were obtained analytically and are indicated in table form in Figure 4.

According to the Ny results corresponding to points 6 and 7, $T_{Ny}^{0.9}$ is the only valid possibility since the Ny for this temperature clearly justifies the expectations about the porosity response of the system in those areas. Thus, $Ny_7$ is above 775 $K^{1/2}\ s^{1/2}\ m^{-1}$ only from $T_{Ny}^{0.9}$ onwards; meanwhile, $Ny_6$ remains below that value up to $T_{Ny}^{0.95}$. The criterion of taking $T_{Ny}^{0.9}$ as the reference to consider Ny in the rest of the points is coherent as well. Thus, the results of points 1 and 2 indicate that these areas will lack porosity, which is coherent with the dimensioning of the feeding system according to the empirical and traditional rules. $Ny_3$ reveals the existence of porosity. The distance of the first riser from the extreme of the bar is larger than the corresponding one according to the empirical rules. The design in this part of the bar was specifically established to evaluate the behavior of an area of the part that it is supposed to solidify last according to the directional solidification principles [25]. In addition, no end effect was expected due to the proximity of the part and the filling system at this extreme of the bar. Nevertheless, the filling system sometimes acts with a riser function as it is filling continuously with hot metal. However, the SF analysis showed that the gate freezes faster than the bar, and some isolated zones appear without the possibility of being fed the corresponding shrinkage, as shown in Figure 2. Consequently, some porosity can appear. Thus, the $Ny_3$ value at 90% of the solidification process is coherent with the expected results, being clearly much lower than 775 $K^{1/2}\ s^{1/2}\ m^{-1}$. Even Ny at 95% is close to that value.

$Ny_5$ indicates the probable presence of porosity just below the riser. This result is not compatible either with the SF criterion or with the traditional empiric criteria, shown in Appendix A. The simulation of the SF established that the riser necking is a hot spot since the SF value is lower than that existing on both sides of the necking area, part and riser. This means that the piece is no longer isolated during the solidification. However, it is possible that the narrow path limited by the necking area is not sufficient to guarantee the transport of enough liquid through the fine interdendritic channels. This statement is confirmed by the results shown in Figure 3, in which it can be seen that the track of connection between the risers and the piece is very narrow. In addition, the thermal gradient is interrupted, and some authors consider that the Niyama criterion is only applicable in limited conditions for directional solidification [30].

With respect to points 4 and 6, inside of the riser, it can be observed that the evolution of the Ny value is clearly more gradual at the beginning of the transition during the solidification, shown in Figure 4d,f, even though, in the case of point 4, the Niyama evolution presents a stepped shape around a value of 500 $K^{1/2}\ s^{1/2}\ m^{-1}$. At this point, the metal arrives colder than at the rest of the riser since this area fills last. In addition, the solidification is faster due to the heat transfer convection of metal–air and conduction existing. All this means that no severe surface shrinkage is expected in comparison to at the top of the sprue, which usually is the last part of the filling system to solidify. Thus, $Ny_4$ is only 5% larger than the critical Ny value, and no conclusive statements can be made for this point. Obviously, $Ny_6$ is very low up to $T_{Ny}^{0.9}$, indicating a severe porosity in this part, as expected. Logically, points 4 and 6 were only employed for testing the results since, technologically speaking, they lack interest.

The methodology applied permitted us to establish a relationship between the Niyama threshold and the temperature at which this should be evaluated in order to obtain results coherent with those expected in the solid fraction analysis and others. In this case, the results obtained for low-carbon steel are totally coherent with those established by other authors. Carlson et al. [16] established the temperature $N_y^{0.9}$ to determine the Niyama value as a criterion for evaluating the porosity in steel casting, which coincides with the value demonstrated herein. All the results obtained are those expected, taking into account that the filling and feeding systems were established according to the classical empirical rules published in the literature [4,5,16,17,23,27,28,31,32]. Thus, very low Ny values were obtained in the area highly distanced from any riser, point 3, and in the inner areas of the risers, point 6, denoting empty volume in those zones; these parts either do not fulfill the classical rules of the feeding system design or take part of their own feeding system, being the last part that solidifies, totally isolated from any liquid part of the casting. In this last case, it reflects that the feeding system works suitably. The areas that solidify sooner, such as those near the end of the bar, point 7, and the surfaces of the risers, point 4, give high Ny values which correspond to zones without porosity. In addition, these areas present the lowest metal temperature at the end of the filling process. Finally, the points covered by the feeding distance but in the middle of the inter-riser length or similar, points 2 and 1, present values of Ny above the threshold one, but close to it, which means that some microporosity can appear.

Figure 5 depicts the results of porosity obtained by FEM simulation using Inspire Cast (Altair Engineering Inc., Troy, MI, USA). The areas shown correspond to those for which the empty volume is at least the value, expressed as a percentage, selected by the arrow existing in the bar legend. The areas involved are coherent with the Niyama analysis previously carried out. Thus, if Figure 5a,b are compared, it can be deduced that there is only a small area in the bar with a percentage of empty volume lower than 20%, which corresponds with the Niyama prediction. No areas with a higher empty volume level than 20% appear in the bar. Only the inner central part of the risers and some very small areas in the riser necking present values of 50%. Obviously, the central part of the risers contains a high level of shrinkage porosity with a large area which is totally empty.

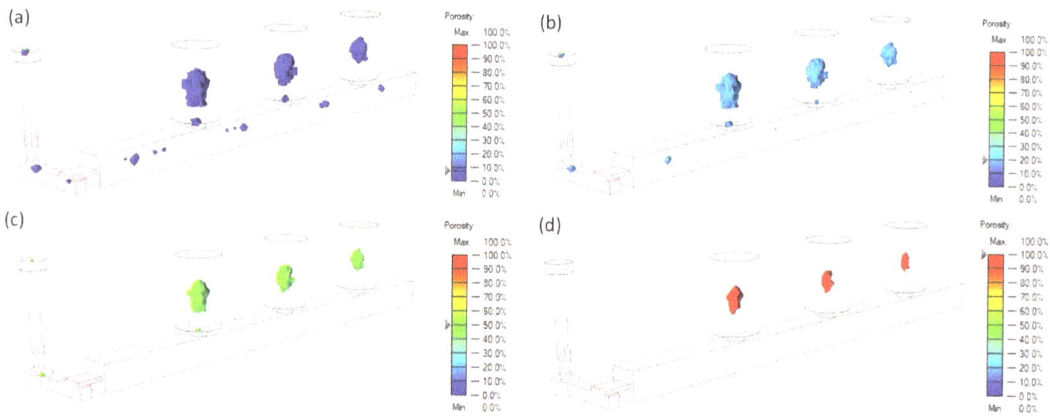


**Figure 5.** Porosity mapping with the areas that present an empty volume of at least (**a**) 10%, (**b**) 20%, (**c**) 50% and (**d**) 100%.

Related to microporosity, this one only appears in punctual areas in the bar, in the parting plane and in the bar surface, as can be appreciated in Figure 6a for a value of 0.04% of empty volume in those areas. Only two small areas with a maximum empty volume of 0.13% can be observed in Figure 6b. In addition, these two areas do not correspond with the points under the Niyama analysis, although they are placed near points 3 and 1, respectively. The microporosity detected can be justified if the compromised Ny values obtained for points away from risers (points 1 and 2) are taken into account. The Ny values of those points are close to the critical established Niyama number, which denotes the possibility of microporosity appearance. Nevertheless, microporosity is not relevant in the system analyzed.

The methodology can be applied to other materials for which the definition of the threshold value is inaccurate and/or for which the temperature to evaluate Ny is not clear [9,14,33–37].

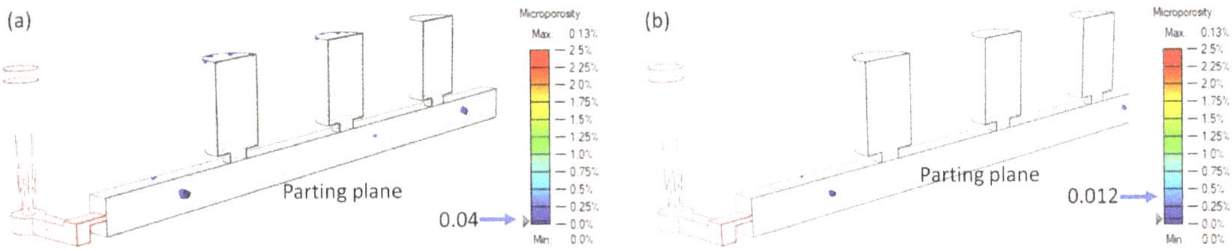


**Figure 6.** Microporosity mapping. (**a**) The microporosity areas contain at least 0.04% of empty volume; (**b**) capture for a level of 0.012% empty areas. The maximum empty volume is 0.13% for only two punctual areas in the bar.

## 5. Conclusions

A methodology to determine the FEM criteria to select the Niyama value in a sand casting process has been established. For that, the design of the system was carried out according to the empirical principles existing in the literature for low-carbon steel casting. Ny evolution during the last solidification of the casting system was analyzed in key areas of the part, along with the solid fraction criterion. The temperature for evaluating the Niyama number should be the one corresponding to 90% of the solidification process. Thus, the results obtained are coherent with those expected and with the results obtained by other authors for the alloy experimented on herein. Thus, very low Ny values were obtained in an area greatly distanced from any riser and in the inner areas of the risers, denoting porosity in those parts; these parts either do not fulfill the classical rules of the feeding system design or take part of their own feeding system. In this last case, it reflects that the feeding system works suitably. The areas that solidify sooner, such as those near the end of the bar, and the surfaces of the risers, give high Ny values that correspond to zones without porosity. Finally, the points covered by the feeding distance present values of Ny above the threshold one, but close to it, establishing that some microporosity can appear.

This methodology may be applied to define other Niyama thresholds and solidification percentages corresponding to other alloys if some experimental or empirical rules are known concerning the casting of those materials.

**Author Contributions:** Conceptualization, M.C.M. and V.M.; methodology, M.C.M. and V.M.; software, M.C.M. and V.M.; validation, M.C.M. and V.M.; formal analysis, M.C.M. and V.M.; investigation, M.C.M. and V.M; resources, M.C.M. and V.M; data curation, M.C.M. and V.M.; writing—original draft preparation, M.C.M.; writing—review and editing, V.M.; visualization, M.C.M. and V.M; supervision, V.M; project administration, M.C.M. and V.M.; funding acquisition, M.C.M. and V.M. All authors have read and agreed to the published version of the manuscript.

**Funding:** This research received no external funding.

**Data Availability Statement:** The data used to support the findings of this study can be shared under request.

**Acknowledgments:** The authors acknowledge the technical assistance of the support team of ALTAIR Spain.

**Conflicts of Interest:** The authors declare no conflict of interest.

## Appendix A

In this appendix, the criteria and rules for designing the feeding system in sand casting of low-carbon steel bars are collected.

The cooling modulus of the risers, $M_r$, is worked out according to Chvorinovs's rule and considering 20% of oversize with respect to the part modulus [37], $M_p$, Equation (A1). $V_r$, $V_p$, $S_r$ and $S_p$ represent the volume and the surface of the riser and the part, respectively.

A height-to-diameter ratio of the risers, $H_r/D_r$, of 1.5 is proposed to guarantee a suitable local metal pressure [4,23].

$$M_r = \frac{V_r}{S_r} = 1.2 \cdot M_p = 1.2\frac{V_p}{S_p} \quad \text{(A1)}$$

The heat transfer surface of the riser is the lateral one, and the top surface, i.e., the bottom surface, is not considered as a cooling path. With respect to the workpiece, for the heat transfer surface calculus, the lateral surface in the gating system on extreme side of the bar is excluded for a similar reason to the one described above.

It is discovered that with three risers the volume criterion for the feeding system, $V_r$, is fulfilled, according to Equation (A2). A riser efficiency, $\xi$, of 14% [4] is considered. The low-carbon steel conductivity coefficient, $\alpha$, is included in Equation (A2).

$$V_r = \frac{\alpha}{\xi}(V_r + V_p) \quad \text{(A2)}$$

The riser neck is designed taking into account the considerations established in Equation (A3) for the height, $h_n$, and the diameter, $D_n$, as a function of the dimensions of the riser $h_r$ and $D_r$, respectively [31].

$$\begin{array}{l} h_n < \frac{D_r}{2} \\ D_n = h_n + 0.2D_r \end{array} \quad \text{(A3)}$$

The location of the risers is set taking into consideration the empirical criteria existing in the literature [4,23–25,28,32]. Taking into account these considerations, the risers are distributed in the system in order to guarantee the feeding action, $L_r$, through the piece. Particularly, the Johnson and Loper rule is fulfilled, Equation (A4). In addition, an end effect, EZL, at the extreme of the bar is considered according to [28] with a ratio width/thickness equals 1 for the workpiece.

$$L_r(\text{mm}) = 80 \cdot \sqrt{M} - 84 \quad \text{(A4)}$$

## References


1. Hardin, R.A.; Beckermann, C. Effect of Porosity on Deformation, Damage, and Fracture of Cast Steel. *Metall. Mater. Trans. A* **2013**, *44*, 5316–5332. [CrossRef]
2. Hardin, R.A.; Beckermann, C. Effect of Porosity on the Stiffness of Cast Steel. *Metall. Mater. Trans. A* **2007**, *38*, 2992–3006. [CrossRef]
3. Hardin, R.A.; Beckermann, C. Prediction of the Fatigue Life of Cast Steel Containing Shrinkage Porosity. *Metall. Mater. Trans. A* **2009**, *40*, 581–597. [CrossRef]
4. Campbell, J. *Complete Casting Handbook: Metal Casting Processes, Metallurgy, Techniques and Design*, 2nd ed.; Butterworth-Heinemann: Oxford, UK, 2015; ISBN 9780444635099.
5. Miguel-Eguía, V.; Manjabacas-Tendero, M.C.; Medina-Ríos, N. Prediction of the Effectiveness of the Feeding System of Carbon Steel Sand Castings Using the Solid Fraction Criterion. Application to Square Bars. *Int. J. Mater Product Tech.* **2016**, *53*, 28–41. [CrossRef]
6. Tavakoli, R. On the Prediction of Shrinkage Defects by Thermal Criterion Functions. *Int. J. Adv. Manuf. Tech.* **2014**, *74*, 569–579. [CrossRef]
7. Lourenço, A.R. Numerical Modelling of Shrinkage Defects Induced by the Feeding Flow in Aluminium Castings. Ph.D. Thesis, University of Gent, Gent, Belgium, 2007. Available online: https://repositorio-aberto.up.pt/bitstream/10216/122834/2/358133.pdf (accessed on 25 August 2023).
8. Niyama, E.; Uchida, T.; Morikawa, M.; Saito, S. A Method of Shrinkage Prediction and its Application to Steel Casting Practice. *Int. Cast Met. J.* **1982**, *7*, 52–63.
9. Khaled, I. Prediction of Shrinkage Porosity in Ti-46Al-8Nb Tilt-Casting Using the Niyama Criterion Function. *Int. J. Metalcast.* **2013**, *7*, 35–42. [CrossRef]
10. Ivanina, E.S.; Monastyrskiy, V.P.; Ershov, M.Y. Quantitative Estimation of Formation of Shrinkage Porosity by the Niyama Criterion. *Inorg. Mater. Appl. Res.* **2022**, *13*, 100–105. [CrossRef]
11. Abootorabi, A.; Korojy, B.; Jabbareh, M.A. Effect of Mould Design on the Niyama Criteria during Solidification of CH3C 80t Ingot. *Ironmak. Steelmak.* **2020**, *47*, 722–730. [CrossRef]

12. Carlson, K.D.; Beckermann, C. Prediction of Shrinkage Pore Volume Fraction Using a Dimensionless Niyama Criterion. *Metall. Mater. Trans. A Phys. Metall. Mater. Sci.* **2009**, *40*, 163–175. [CrossRef]
13. Dwulat, R.; Janerka, K. Evaluation of the Metallurgical Quality of Nodular Cast Iron in the Production Conditions of a Foundry. *J. Manuf. Mater. Process.* **2023**, *7*, 18. [CrossRef]
14. Ignaszak, Z. Discussion on Usability of the Niyama Criterion for Porosity Predicting in Cast Iron Castings. *Arch. Foundry Eng.* **2017**, *17*, 196–204. [CrossRef]
15. Díez-Rodríguez, D. Gradient Based Porosity Calculation in Casting Simulation. Master´s. Thesis, Poli-technical University of Catalunya, Barcelona, Spain, 2017. Available online: https://upcommons.upc.edu/bitstream/handle/2117/113965/DanielDiez-TFM.pdf (accessed on 25 August 2023).
16. Carlson, K.D.; Ou, S.; Hardin, R.A.; Beckermann, C. Development of New Feeding-Distance Rules Using Casting Simulation: Part I. Methodology. *Metall. Mater. Trans. B* **2002**, *33*, 731–740. [CrossRef]
17. Ou, S.; Carlson, K.D.; Hardin, R.A.; Beckermann, C. Development of New Feeding-Distance Rules Using Casting Simulation: Part II. The New Rules. *Metall. Mater. Trans. B* **2002**, *33*, 741–755. [CrossRef]
18. Jain, N.; Carlson, K.D.; Beckermann, C. Round Robin Study to Assess Variations in Casting Simulation Niyama Criterion Predictions. In Proceedings of the 61st SFSA Technical and Operating Conference; paper 5.5.; Steel Founders' Society of America, Chicago, IL, USA, 2007. Available online: https://user.engineering.uiowa.edu/~becker/documents.dir/sfsa/2007-5.5_niyama.pdf (accessed on 12 September 2023).
19. Ding, H.; Jin, X.; Chen, T.; Li, H. Study on Casting Defect Control of Austenitic 304 Complex Structural Parts. *Int. J. Met.* **2023**, *17*, 1427–1438. [CrossRef]
20. Wang, T.; Zhou, J.; Wang, L.; Zhang, Z.; Zhang, H.; Xia, X. Research and Development of Gating and Riser Process Optimization Integrated System for Steel Casting. *Int. J. Met.* **2023**, *17*, 2452–2468. [CrossRef]
21. You, L.; Yao, L.; Li, X.; Jia, G.; Lv, G. Numerical Simulation and Casting Process Optimization of Cast Steel Node. *International J. Adv. Manuf. Tech.* **2023**, *126*, 5215–5225. [CrossRef]
22. Liu, J.; Yang, L.; Fang, X.; Li, B.; Yang, Y.; Fang, L.; Hu, Z. Numerical Simulation and Optimization of Shell Mould Casting Process for Leaf Spring Bracket. *China Foundry* **2020**, *17*, 35–41. [CrossRef]
23. Beeley, P. *Foundry Technology*, 2nd ed.; Butterworth-Heinemann: Madras, India, 2001.
24. Jain, P.L. *Principles of Foundry Technology*, 5th ed.; Tata McGraw-Hill: New Delhi, India, 2009.
25. Wlodawer, R. *Directional Solidification of Steel Castings*; Pergamon Press: Oxford, UK, 1966.
26. Campbell, J. *Castings*, 2nd ed.; Elsevier Butterworth-Heinemann: Cornwall, UK, 2003.
27. Campbell, J. *Castings Practice. The 10 Rules of Castings*; Elsevier Buttherworth-Heinemann: King's Lynn, UK, 2004.
28. Steel Founders' Society of America. *Feeding & Risering Guidelines for Steel Castings*; Steel Founders' Society of America: Crystal Lake, IL, USA, 2001.
29. *Altair, Altair Inspire Cast Version 2022.1.1*; Altair Engineering Inc.: Troy, MI, USA, 2022.
30. Spittle, J.A.; Brown, S.G.R.; Sullivan, J.G. Application of Criteria Functions to the Prediction of Microporosity Levels in Castings. In Proceedings of the Fourth Decennial International Conference on Solidification Processing, Sheffield, UK, 7–10 July 1997; pp. 251–255.
31. Viswanathan, S.; Apelian, D.; Donahue R., J.; DasGupta, B.; Gywn, M.; Jorstad, J.L.; Monroe, R.W.; Sahoo, M.; Prucha, T.E.; Twarog, D. *ASM Handbook*, 9th ed.; ASM International: Materials Park, OH, USA, 2008; Volume 15.
32. Pellini, W.S. Factors which Determine Riser Adequacy and Feeding Range. *Am. Foundrymen'S Soc. Trans.* **1953**, *61*, 61–80.
33. Li, J.; Chen, R.; Ma, Y.; Ke, W. Characterization and Prediction of Microporosity Defect in Sand Cast WE54 Alloy Castings. *J. Mater. Sci. Technol.* **2014**, *30*, 991–997. [CrossRef]
34. Carlson, K.D.; Beckermann, C. Use of the Niyama Criterion to Predict Shrinkage-Related Leaks in High-Nickel Steel and Nickel-Based Alloy Castings. In Proceedings of the 62nd SFSA Technical and Operating Conference; Steel Founders' Society of America, Crystal Lake, IL, USA, 2008. Available online: https://user.engineering.uiowa.edu/~becker/documents.dir/sfsa/2008-5.6_nileakers.pdf (accessed on 10 August 2023).
35. Kang, M.; Gao, H.; Wang, J.; Ling, L.; Sun, B. Prediction of Microporosity in Complex Thin-Wall Castings with the Dimensionless Niyama Criterion. *Materials* **2013**, *6*, 1789–1802. [CrossRef] [PubMed]
36. Liotti, E.; Previtali, B. Study of the Validity of the Niyama Criteria Function Applied to the Alloy AlSi7Mg. *La Metall. Ital.* **2006**, *9*, 33–37.
37. Jeancolas, M. Donnés Générales Sur Le Masselottage Des Piéces Molées En Sable. *Fonderie* **1961**, *341*, 43–56.

metals

MDPI

Article

# Study on Simulation of Mold Filling and Solidification Characteristics of Hypereutectic High-Chromium Cast Iron by Lost Foam Suspension Casting

**Hongliang Ma and Hanguang Fu ***

Key Laboratory of Advanced Functional Materials, Ministry of Education, Department of Materials Science and Engineering, Beijing University of Technology, Beijing 100124, China; mahl@emails.bjut.edu.cn
* Correspondence: hgfu@bjut.edu.cn; Tel.: +86-10-67396093; Fax: +86-10-67396244

**Abstract:** In this paper, the finite element software ProCAST version 14.5 was used to simulate the temperature field, flow field and defect prediction in the filling and solidification process of hypereutectic high-chromium cast iron. The effects of pouring temperature, negative pressure and the amount of suspension agent added during the technological process were explored. The optimum process parameters were presented. It was found that the suspension agent has a certain hindrance to the filling process, but the filling process remains stable. In the solidification stage, 89.4% of the suspension agent melted, resulting in a relative supercooling degree of 50 °C, which had a certain chilling effect and improved the solidification rate.

**Keywords:** lost foam casting; suspension casting; numerical simulation; solidification characteristics

## 1. Introduction

Hypereutectic high-chromium cast iron contains a large number of M7C3 primary carbides which are supported and protected by the matrix so that it has better wear resistance than traditional eutectic and hypoeutectic alloys. During the solidification process of hypereutectic high-chromium cast iron, coarse primary carbides are formed, which reduces toughness. Defects such as cracks and shrinkage holes also easily occur during casting production, which restricts its application and promotion to a certain extent [1]. In order to improve the toughness of hypereutectic high-chromium cast iron and improve its comprehensive mechanical properties, domestic and foreign researchers have done a lot of research on refining primary carbides; the main methods include metamorphism and inoculation [2–5], heat treatment [6–10], pulsed current [11,12] treatment, etc. Previous studies on refining primary carbides of hypereutectic high-chromium cast iron have also been carried out by our research group. The studies have shown that appropriate modification treatment (RE-Mg), element regulation and heat treatment are helpful in refining primary carbides. It has been found that refining primary carbide combined with heat treatment and alloying can improve the corrosion and wear resistance of hypereutectic high-chromium cast iron [13].

Suspension casting technology (also known as micro-cooling) is a new casting method proposed by Soviet experts in the 1960s. This is a casting method in which a certain amount of metal powder is added to the metal flow as a suspension agent when pouring liquid metal into the mold [14]. The suspension agent is extremely fine and plays an internal chilling role, while the alloying elements contained in it play a metamorphic role, promoting heterogeneous nucleation and thus refining the crystallization structure, improving metal density, reducing casting defects such as shrinkage porosity, shrinkage porosity, segregation and hot cracking and improving mechanical properties [15]. Suspension casting technology is widely used in a variety of materials such as aluminum matrix composites [16] like cast steel, cast iron, etc. to improve the hardness, wear resistance and corrosion resistance of castings and to refine the microstructure. The important process parameters of suspension casting include the composition and quantity of the suspension agent and the size of the particles. Li Qiushu et al. [17] studied the effect of suspension casting on the microstructure and mechanical properties of high-chromium white cast iron by adding 66%Cr-Fe alloy as a suspension agent using the punching method. The results showed that the microstructure of high-chromium cast iron can be significantly improved by adding the suspension agent (adding 1.5 wt%); the continuous network carbide becomes a fine and uniform granular carbide, the impact toughness is increased by 20–30% and the wear resistance is increased by 20%. Dai Binyu [18] used iron powder as a suspension agent to manufacture a high-chromium cast iron refractory brick mold using the method of suspension casting with the agent added in metal flow. When the amount of suspension agent was 0.5~1 wt%, the microstructure of the high-chromium cast iron was refined, the carbide was changed from a continuous mesh to an intermittent form and the wear-resistant performance and service life were improved. Zhi Xiaohui et al. [19] found that the addition of a 2.1% high carbon steel suspension agent outside the mold could improve the microstructure characteristics of a cast hypereutectic Fe-20Cr-4C alloy and obtain fine and uniform carbides. The average size of the primary carbides was reduced from 30 μm to 5.7 μm, and the impact toughness was improved from 2.5 $J/cm^2$ to 5.5 $J/cm^2$. It can be seen that suspension casting can effectively refine the primary carbides of hypereutectic high-chromium cast iron and improve its toughness and wear resistance.

Lost foam casting (LFC) or expendable pattern casting (EPC), also known as vacuum-evaporative pattern casting, is a nearly unlimited, precise, vacuum seal casting and solid casting process grafted from new technology and representing the developmental direction of casting methods in the 21st century. It is a precision casting method that uses foam to make a solid shape of exactly the same structure and size of the parts. After dipping in refractory adhesive paint and drying, dry sand modeling and vibration compaction is carried out and the liquid metal is poured to make the shape disappear under heat and gasification so as to obtain metal parts with the same shape [20,21]. The important process parameters affecting the microstructure and properties of castings in the EPC process include the pouring temperature and negative pressure. Suyitno et al. [22] studied the relationship between the fluidity, porosity and surface roughness of gray cast iron and pouring temperature in the lost foam casting process. The experimental results show that the fluidity, porosity and surface roughness of gray cast iron increase with an increase in the casting temperature within the temperature range of 1300–1400 °C. The higher the pouring temperature, the lower the viscosity of the liquid metal and the more serious the turbulence phenomenon during pouring. The turbulence makes it easy to enclose the gas into the liquid metal, so the porosity of the casting increases. Xie Mingguo et al. [23] studied the influence of vacuum conditions on the solidification structure of hypoeutectic gray cast iron in lost foam casting and found that the influence of vacuum conditions on the filling process was greater than that on the solidification process, and the microstructure of the thin-wall region of the casting was more sensitive to the influence of vacuum conditions. Wu Guohua et al. [24] studied the effect of vacuum degree on the porosity of aluminum alloy cast by lost foam casting. These studies show that lower negative pressure is beneficial in reducing the porosity of aluminum alloy casting.

In this experiment, the method of preparing composite materials by lost foam casting was used for reference [21,25,26], and the process of suspension casting and lost foam casting was combined. With the help of the foam, the suspension agent could be more evenly distributed in the casting, which could refine the microstructure of cast iron and improve the mechanical properties such as hardness and wear resistance of the material [27–29]. Liu Gensheng et al. [27] studied the effect of adding 89%Cr-Fe as a suspension agent of lost foam casting low chromium white cast iron on its performance and found that the matrix microstructure changed from pearlite to sortenite and tristenite and the pearlite sheet spacing decreased. With the increase of the Cr/C ratio, the carbide type changes from

$M_3C$ to $M_7C_3$, and the wear resistance is the best when the addition amount is 2.0–2.5 wt%, which is 35% higher than that of the conventional lost foam casting process.

At present, although the simulation of the casting filling and solidification process using large-scale commercial software such as ProCAST has been relatively mature [30–34], the lost foam suspension casting technology has been widely used in the development and production of various types of wear-resistant and heat-resistant iron and steel castings and has achieved good results [28,29]. However, there is no report on the numerical simulation of lost foam suspension casting. In this experiment, the numerical simulation method was introduced into the research process of lost foam suspension casting and the filling and solidification characteristics of hypereutectic high-chromium cast iron were simulated and analyzed, providing reference data for the popularization and application of hypereutectic high-chromium cast iron.

## 2. Mathematical Model

The flow process of casting metal filling includes the energy loss of the liquid metal itself and the flow and solidification process of the liquid metal under non-constant temperature conditions [35,36]. The filling flow process of liquid metal follows three conservation laws: mass conservation, momentum conservation and energy conservation. The continuity equation, momentum equation (Navier–Stokes equation), volume function equation and the energy equation are usually used to describe this process, as shown in Formulas (1)–(4). The actual pouring process of liquid metal is very complicated because of the complex and diverse design structure of the pouring system. The turbulent flow factors of liquid metal make the filling mostly turbulent, so the calculation of turbulent flow treatment needs to be introduced in the simulation process. At present, the numerical simulation of the casting mold filling process mostly adopts the SOLA-VOF method. SOLA is an iterative method to solve the velocity field-pressure field, and VOF is a method to deal with free surfaces (short for volume function), as shown in Formulas (5)–(7).

Mass conservation equation:

$$D = \frac{\partial u}{\partial x} + \frac{\partial v}{\partial y} + \frac{\partial w}{\partial z} = 0 \tag{1}$$

Momentum conservation equation (N–S equation):

$$\rho\left(\frac{\partial u}{\partial t} + u\frac{\partial u}{\partial x} + v\frac{\partial u}{\partial y} + w\frac{\partial u}{\partial z}\right) = -\frac{\partial p}{\partial x} + \rho g_x + \mu\nabla^2 u \tag{2}$$

$$\rho\left(\frac{\partial v}{\partial t} + u\frac{\partial v}{\partial x} + v\frac{\partial v}{\partial y} + w\frac{\partial v}{\partial z}\right) = -\frac{\partial p}{\partial x} + \rho g_y + \mu\nabla^2 v \tag{3}$$

$$\rho\left(\frac{\partial w}{\partial t} + u\frac{\partial w}{\partial x} + v\frac{\partial w}{\partial y} + w\frac{\partial w}{\partial z}\right) = -\frac{\partial p}{\partial x} + \rho g_z + \mu\nabla^2 w \tag{4}$$

Here, D is the degree of dispersion and u, v and w are the components (m/s) of the velocity vector in the X, Y and Z directions of the coordinate system. P is the pressure per unit density (Pa). $\mu$ is the kinematic viscosity ($m^2/s$). g is the gravitational acceleration ($m/s^2$). $\nabla^2$ is the Laplace operator. $\rho$ is the metal fluid density ($kg/m^3$).

Energy equation:

$$\rho c\frac{\partial T}{\partial t} + \rho cu\frac{\partial T}{\partial x} + \rho cv\frac{\partial T}{\partial y} + \rho cw\frac{\partial T}{\partial z} = \frac{\partial}{\partial x}\left(k\frac{\partial T}{\partial x}\right) + \frac{\partial}{\partial y}\left(k\frac{\partial T}{\partial y}\right) + \frac{\partial}{\partial z}\left(k\frac{\partial T}{\partial z}\right) + S \tag{5}$$

When the volume function method is used to track the movement of the free surface, it is also necessary to solve the F-function equation. When the F-value is one, the grid is in a full state and there is no free surface; when the F-value is zero, the grid is empty and there is no free surface; when the F-value is between zero and one, the grid is semi-full and there is a free interface in the grid. F function equation:

$$\frac{\partial f}{\partial t}+u\frac{\partial f}{\partial x}+v\frac{\partial f}{\partial y}+w\frac{\partial f}{\partial z}=0 \quad (6)$$

The reasonable amount of suspension agent added in suspension casting can be calculated according to the following formula [37]:

$$S=100\bullet\frac{t_1\left(1-0.0144\frac{C\sqrt{t_1}}{R}\right)-t_2+39}{t_2} \quad (7)$$

Here, $t_1$ is the pouring temperature and $t_2$ is the liquidus temperature.

## 3. Experimental Method

### *3.1. Three-Dimensional Modeling and Mesh Division*

In order to reduce the calculation time and improve the calculation efficiency in the process of studying the characteristics of the lost foam suspension casting technology by using the simulation method, a simple cuboid ingot was selected as the experimental research object. Creo 4.0 software was used to model castings, suspensions and pouring systems. The suspensions were distributed evenly in cuboid castings with a metal ball 2 mm in diameter. Both the size of the suspension agent and the size of the mesh division are very important. In actual production, the size of the suspension agent is usually between 0.5–2.5 mm. If the size is too small, a finer mesh will be needed in mesh division to slow down the calculation speed; if the size is too large, the performance improvement effect in actual production will be poor. The size of the casting is 100 mm × 50 mm × 32.5 mm, and the size of the mold is 140 mm × 125 mm × 58 mm. The bottom casting system design is adopted. The assembly is imported into the MeshCAST module in ProCAST 14.5 software for verification, generation and repair of the surface mesh and generation and inspection of the body mesh. The meshing model is shown in Figure 1, in which there is a transitional area of mesh size between the body mesh of the suspension agent and the casting, as shown in Figure 2.

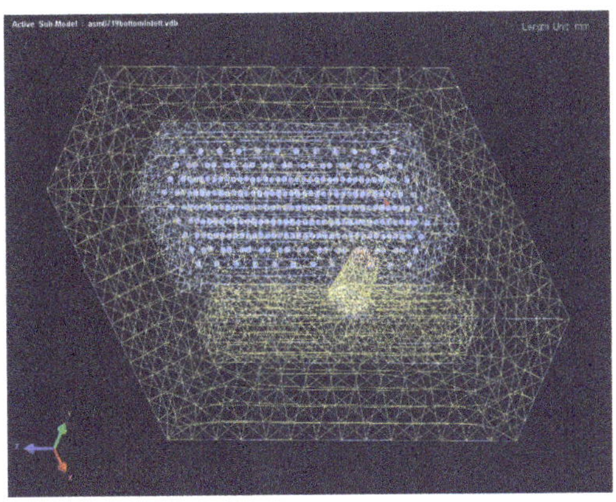

**Figure 1.** Casting mesh division model.

**Figure 2.** Casting and suspension mesh transition diagram.

### 3.2. Material Selection

In this experiment, 35Cr-4C hypereutectic high-chromium cast iron was used, which has high hardness and excellent wear resistance. The chemical composition of hypereutectic high-chromium cast iron is shown in Table 1 and its material properties were calculated by ProCAST.

**Table 1.** Chemical composition of hypereutectic high-chromium cast iron (mass fraction, wt%).

| C | Cr | Mo | Mn | Si | Fe |
|---|---|---|---|---|---|
| 4 | 35 | 1 | 0.6 | 0.8 | Bal. |

### 3.3. Simulation Parameter Settings

Simulation parameters are the basis for the numerical simulation process, and the reliability of the simulation results depends on the rationality of the selection of simulation parameters. ProCAST 14.5 software was innovatively used in this experiment to study the process characteristics of the lost foam suspension casting, so the setting of simulation parameters and boundary conditions was very important. The key simulation parameters in lost foam suspension casting included pouring temperature, pouring speed, negative pressure, type of suspension agent, addition amount of suspension agent and particle size of suspension agent. This experiment focused on the influence of pouring temperature, negative pressure and the amount of suspension agent on the process and process defects.

The initial and boundary conditions were set as follows: The liquid metal temperature (that is, the initial temperature set in the pouring cup) was the pouring temperature (1480 °C, 1540 °C and 1600 °C). TiFe particles with a diameter of 2 mm were used as the suspension agent. The casting material was sand permeable foam specifcally for lost mold casting in ProCAST. The casting and runner were made of foam material and the sprue cup was set as alloy. The initial temperature of the foam material, mold and suspension agent should be 25 °C. The heat transfer coefficient between the mold and the liquid metal was 400 W/m$^2$·K, and the heat transfer coefficient between the liquid metal and the suspension agent was 2000 W/m$^2$·K. In the actual casting process, the heat transfer coefficient should vary with time, but in this experiment, the heat transfer coefficient was simplified to a constant value for the convenience of the research. The outside of the mold and gate were set to air cooling conditions, and the negative pressure was set by creating a pressure difference between the outside of the mold and the outside of the gate. The negative pressure values could be 0.04 MPa and 0.06 MPa. The casting speed was set to 2 kg/s.

It is necessary to pay attention to the special situation of interface heat transfer setting in the simulation parameter setting of lost die casting. EQUIV (equivalent type) is usually used in the heat exchange process between two identical materials, where the two interface units share a common node between the interface and there is a continuous temperature field at the interface. In this experiment, although the casting system and casting were set as alloy and foam respectively, with different material types, in order to ensure that the real situation of heat exchange between liquid metal and foam materials could be simulated, the interface between the two was usually set as EQUIV type, so that the temperature field during simulation was continuous.

## 4. Results and Discussions

### *4.1. Shrinkage Porosity Simulation*

The solidus method, critical solid fraction method, temperature gradient method and Niyama($G/\sqrt{R}$)method were proposed to effectively predict casting defects generated during the solidification [38]. In this experiment, the Niyama($G/\sqrt{R}$) method of ProCAST software was used to predict the shrinkage and porosity in castings. By comparing the results of different negative pressure, casting temperature and the amount of suspension agent, the optimal process parameters were selected for the subsequent analysis of the filling and solidification process.

The role of vacuum-negative pressure is to fix the dry sand in the process of pouring and solidification and to form an exhaust cycle, so that the gas generated by foam pyrolysis and gasification during the pouring process is discharged to the outside of the mold, thereby reducing the casting defects. Zhao Zewen et al. [39] found through an orthogonal experiment that in the lost foam casting process of cast iron with a bottom casting system, the negative pressure had a greater impact on the shrinkage and porosity results than the pouring temperature. According to this conclusion, the influence of negative pressure on the results of porosity and porosity reduction was given priority in this experiment.

In this experiment, the results of shrinkage porosity under the negative pressure of 0.04 MPa and 0.06 MPa were compared, as shown in Figure 3a,b. It can be found from the results that a certain size of porosity and shrinkage holes will be produced at each suspension agent particle, but these porosity and shrinkage holes are very small or even negligible in the actual production. It is speculated that the large shrinkage and porosity of the suspension agent in the simulation software is because the simulation software cannot take into account the exchange of metal materials in different regions. Without considering the shrinkage porosity at the suspension agent, the shrinkage porosity of the casting with a vacuum degree of 0.06 MPa (0.0008 $cm^3$) is significantly smaller than that of the casting with a vacuum degree of 0.04 MPa (0.0085 $cm^3$). Appropriately increasing the negative pressure (from 0.02 MPa to 0.05 MPa) can effectively eliminate dimensional deformation, box collapse, slag inclusion, sand fusion and other defects of castings, shorten casting time and improve casting quality [40].

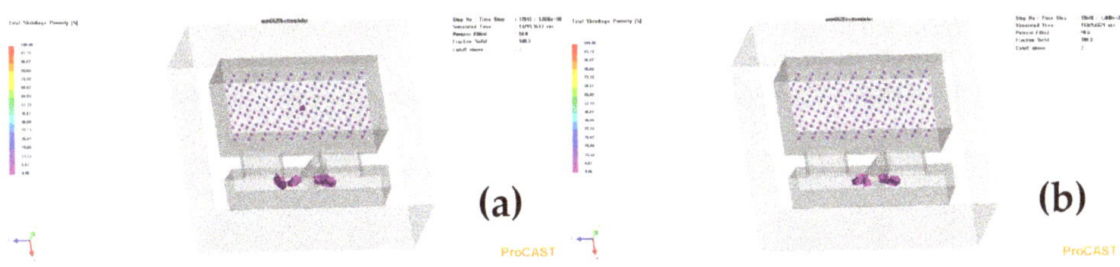


**Figure 3.** Effect of negative pressure on shrinkage porosity of castings: (**a**) negative pressure is 0.04 MPa; (**b**) negative pressure is 0.06 MPa.

The pouring temperature of the lost foam suspension casting will also affect the shrinkage and porosity results. Appropriately increasing the pouring temperature will help improve the liquid metal flow, but too high a pouring temperature will increase the liquid shrinkage of the casting and the volume of shrinkage and porosity formed inside the casting [41]. The casting temperature of hypereutectic high-chromium cast iron is generally about 1400 °C, considering that the casting temperature of lost foam castings is generally about 80 °C higher than that of ordinary sand casting, and the presence of a suspension agent will hinder the flow of the liquid metal, so 1480 °C, 1540 °C and 1600 °C were selected in this experiment. When other conditions are the same, the shrinkage porosity is the lowest when the pouring temperature is 1540 °C.

In this experiment, 0.5%, 0.75% and 1% of suspension additives were selected for discussion. By comparing the results of shrinkage porosity, it was found that in the range of 0.5–1%, the larger the amount of suspension agent added, the lower the shrinkage porosity. When the amount of suspension agent added reaches 1%, the shrinkage porosity reaches $3 \times 10^{-5}$ cm$^3$, which can be said that the shrinkage porosity of simple cuboid castings is almost eliminated.

In summary, the optimal process parameters within the scope of this experiment are a negative pressure of 0.06 MPa, a pouring temperature of 1540 °C and a suspension additive amount of 1%, so subsequent studies are based on the model of optimal process parameters.

### *4.2. Filling Characteristics*

The filling process of lost foam suspension casting is similar to that of lost foam casting. When no suspension agent is added, the casting filling time is 0.87 s, and when the addition amount of the suspension agent is 1%, the casting filling time is 0.95 s. Since the suspension agent almost does not melt during the filling process (as shown in Figure 4), the suspension agent plays a certain role in obstructing the flow of the liquid metal, so the time is longer.

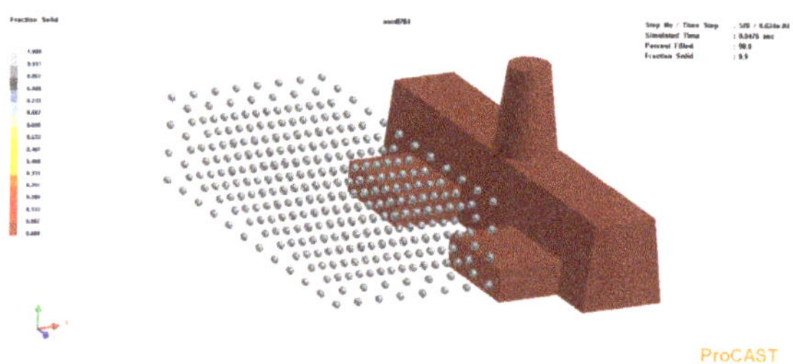


**Figure 4.** Results of solid phase ratio of suspension agent at the end of filling stage.

After encountering high-temperature liquid metal, the foam decomposes and vaporizes, forming an air gap layer between the foam and the liquid metal. Researchers have used the cold quenching method, high-speed photography method and the electrode contact method to observe the flow patterns of different castings in lost foam casting [36,42]. For lost foam casting, the pyrolysis process of foam is the control link of the casting mold filling process and its essence is the reverse pressure resistance of pyrolysis gas to the liquid metal flow [43]. As a result, the filling process of the lost foam casting is different from that of ordinary sand casting. The liquid metal usually enters the casting cavity from the inner runner and fills forward in a radial arc, replacing the foam pattern layer by layer, and finally filling to the farthest part from the inner gate [44]. Under the parameters of this experiment, the thickness of the air gap layer is about 0.25 cm, and the gas pressure of the air gap layer is about 4.5 bar. During the filling process of the liquid metal, the foam stably advances until the filling is complete, as shown in Figure 5. It can be observed through the movement process of the air gap layer that the liquid metal filling is smooth and advances from the gate in a circular arc with decreasing curvature. The presence of the suspending agent hinders the liquid metal filling, but due to the uniform distribution of the suspending agent in the cavity of the casting, there is no obvious influence on the overall shape of the liquid metal filling. The filling process is smooth until the end, and there is no backflow caused by the impact of liquid metal casting, which can effectively avoid the uneven distribution of the suspending agent in the actual production process.

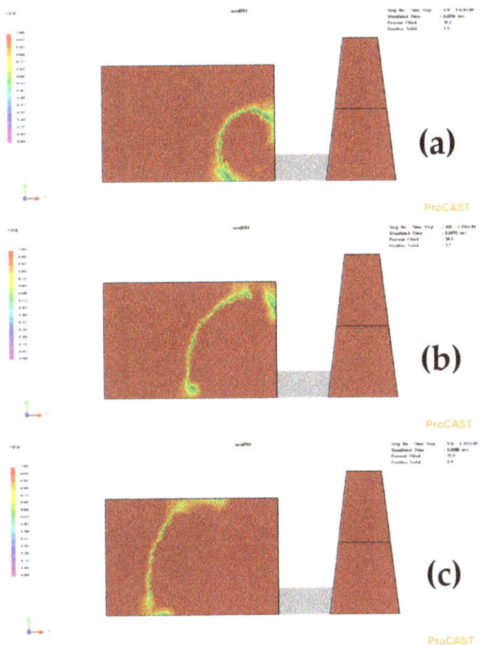


**Figure 5.** Advancing process of air gap layer during mold filling: (**a**) mold filling to 35%; (**b**) mold filling to 50%; (**c**) mold filling to 70%.

From the temperature field results of the overall observation of this model filling process, as shown in Figure 6, there is a color fluctuation of the flowing metal front, which indicates that the high-temperature liquid metal is pyrolyzing the foam material. Although the temperature of the liquid metal front of the filling process fluctuates, it has always had a high temperature and good fluidity, so that the foam pattern can be fully decomposed, no casting defects such as slag inclusion and wrinkled skin are produced and the casting can be smoothly formed. The filling time of the casting to 30% is 0.81 s. It takes 0.86 s for the filling to reach 60% and the filling ends at 0.95 s. The overall filling process is slow at first, then fast and then slow, which also conforms to the filling characteristics of the bottom pouring system [39,45]. The reason for the slow start is that at the beginning of pouring, the temperature of the liquid metal is high and the air volume is large at the moment of foam decomposition, but the exhaust capacity of dry sand is limited, resulting in an increase in the back pressure of gas accumulation at the front end of the filling mold, which further affects the filling mold. At the initial stage of filling, the high temperature liquid metal drops rapidly after encountering the suspension agent, which improves the excessive foam gas generated by the high temperature liquid metal to a certain extent and improves the filling ability of the liquid metal. However, after the temperature of the liquid metal is reduced, the viscosity will also have a certain increase, and the filling ability will be weak from this point of view. Therefore, the addition of the suspension agent did not change the phenomenon of slow filling in the early stage. In the subsequent filling stage, the chilling effect of the suspension agent, the gas generation of the foam and the exhaust effect of the dry sand gradually reached a balance. With the decrease of liquid metal temperature, the filling capacity was improved. The final stage of the filling time was longer because the liquid metal reached the place that needs to be filled far away and the heat was lost, making the filling ability weak. The results show that the filling flow was stable and spatter-free.

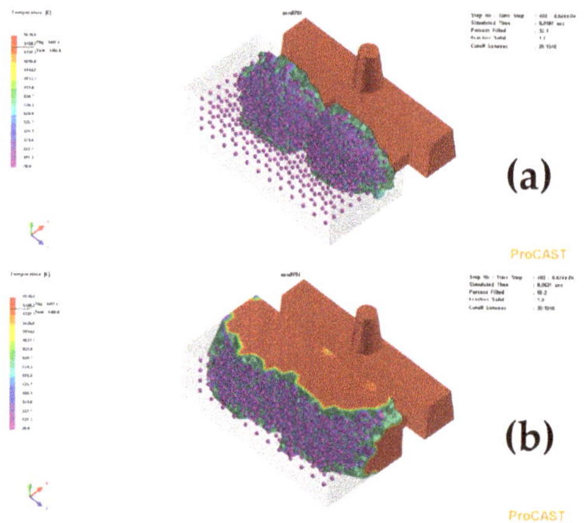


**Figure 6.** Temperature field results of filling process: (**a**) mold filling to 30%; (**b**) mold filling to 60%.

As shown in Figure 7, the Voids result shows that no foam material is in suspension surrounded by the liquid metal, the material is decomposed completely and can be discharged in time and there is no residual foam, so no defects such as slag inclusion and porosity will be formed.

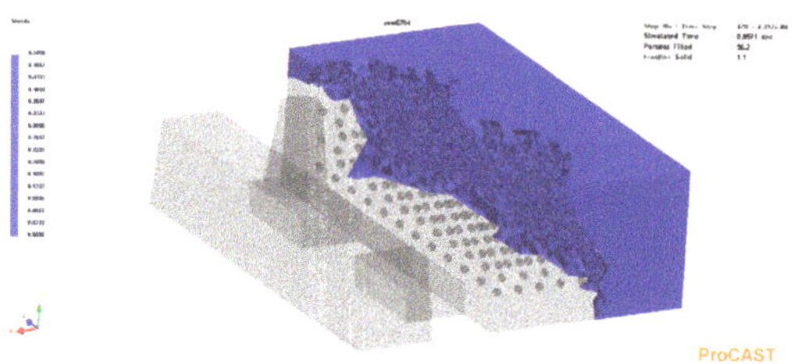


**Figure 7.** Voids result in the mold-filling process.

The temperature curves of the filling and solidification process at the entrance of the inner gate (Point 1) and the farthest part of the casting from the gate, that is, the last part of the theoretical filling (Point 2), are respectively shown. The schematic diagram of the point location is shown in Figure 8, and the temperature curve is shown in Figure 9. Compared with the temperature curves of the two points, the liquid metal quickly rises to the highest temperature after filling to the position, goes through the rapid cooling stage, enters the latent heat release period and finally enters the stable cooling period. Point 1 is heated to the highest temperature by 0.14 s, and Point 2 is heated to the highest temperature by 0.05 s. Since Point 1 is at the lower side of the gate, high-temperature liquid metal will continue to enter the casting, a large number of bubbles will crack, burn and constantly absorb heat, resulting in temperature fluctuations at the liquid metal front and thus a relatively slow temperature rise. As for Point 2, the foam pattern has been completely replaced by liquid metal, the filling process ends, the front flow of liquid metal disappears and only the internal flow of liquid metal exists. Because of the local energy imbalance, the temperature fluctuation is small at this time and the temperature rises faster. In this experiment, due to the small size of the casting and the short filling time, the liquid metal flows from Point 1 to Point 2 very quickly after entering the cavity, so there is only a slight difference in the cooling curve.

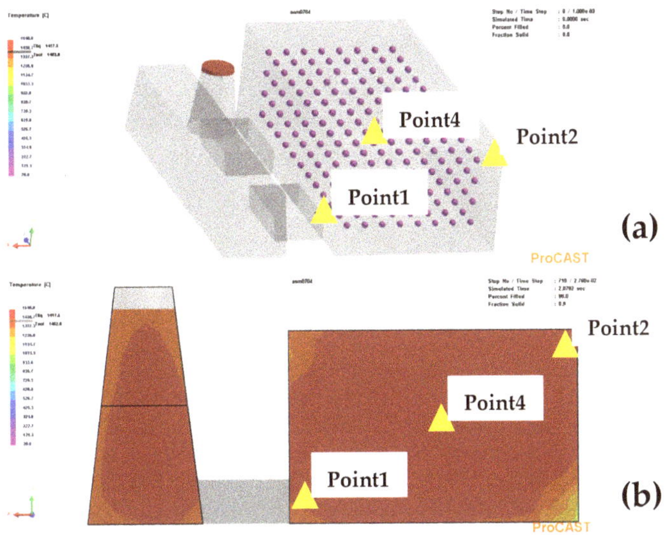


**Figure 8.** Schematic diagrams of the points taken at the entrance of the inner gate (Point 1), at the furthest point of the casting from the inner gate (Point 2) and at the center of the casting (Point 4): (**a**) three-dimensional schematic diagram of the location of the points taken; (**b**) schematic diagram of the location of the points taken in the XY-plane.

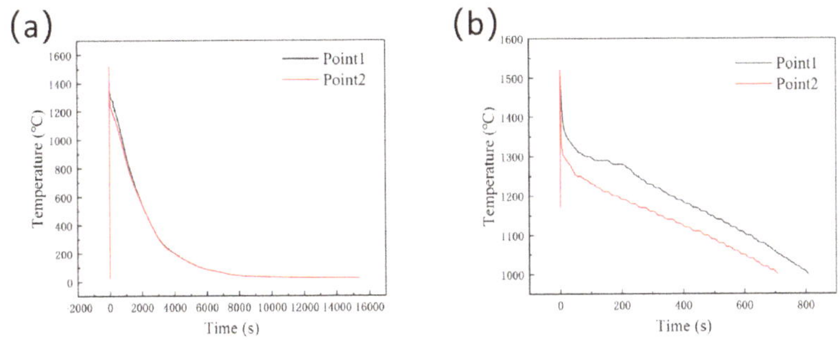


**Figure 9.** Temperature profiles at the entrance of the inner gate (Point 1) and at the furthest point of the casting from the inner gate (Point 2): (**a**) overall temperature results; (**b**) temperature profiles near high-temperature range.

### *4.3. Solidification Characteristics*

The suspension agent keeps heating up and melting in the solidification stage of the casting and most of the suspension agent usually melts in the actual production process of suspension casting. The Cut-off function was used to select the part whose temperature was lower than the liquidus temperature of TiFe (1417 °C), as shown in Figure 10. There were 388 metal spheres in the suspension, of which 41 were not completely melted. That is, 89.4% of the suspension was completely melted. It can be seen that the unmelted part was concentrated at the edge of the casting. Due to the long filling distance of the liquid metal and the short heating time of the suspension agent, the phenomenon of incomplete melting was caused. At the same time, the diameter of 2 mm was relatively large, and the suspension agent was not easy to melt to a certain extent.

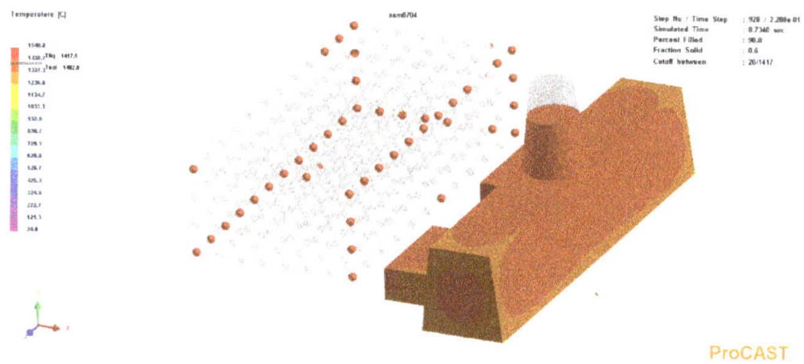


**Figure 10.** Unmelted part of the suspension agent.

By comparing the temperature field results of the model without the suspending agent and the model with 1% suspending agent, it was found that the chilling effect of the suspending agent exists but is not obvious. In this experiment, the geometric center position of the casting was selected as the temperature measurement point (Point 3 and Point 4) to compare the effect of the suspension agent. The location of the casting point (Point 3) without adding the suspension agent is shown in Figure 11 and the temperature curve at Point 3 and Point 4 is shown in Figure 12. It can be found from the temperature curve that the chilling effect of the suspension agent occurs within 0.44 s, the heat absorption time is very short and the maximum temperature drop is 50 °C. In addition, we studied the temperature field around the suspension agent. As shown in Figure 13, the suspension agent had an obvious heat exchange with the surrounding liquid metal during the heating and endothermic process, thus achieving the chilling effect. Zhi Xiaohui et al. [19] studied the microstructural refinement of a cast hypereutectic $Fe_{20}Cr_4C$ alloy using a fluctuation method (suspension casting). With the treatment of 0.7% fluctuation, there is a maximum temperature drop of about 37 °C. This experimental result is in general agreement with the conclusion of the simulation. It is speculated that the cooling effect of the suspension agent is not obvious due to the large particle size, small distribution density and relatively insufficient addition amount. If the amount of suspension agent is increased, the temperature of the alloy can be further reduced, the solidification rate can be accelerated, the nucleation rate of the alloy can be increased and the growth of the primary carbide can be hindered. In addition, any powder that is not fully melted can serve as the heterogeneous nuclear point of the primary carbide, resulting in the refinement of the final carbide.

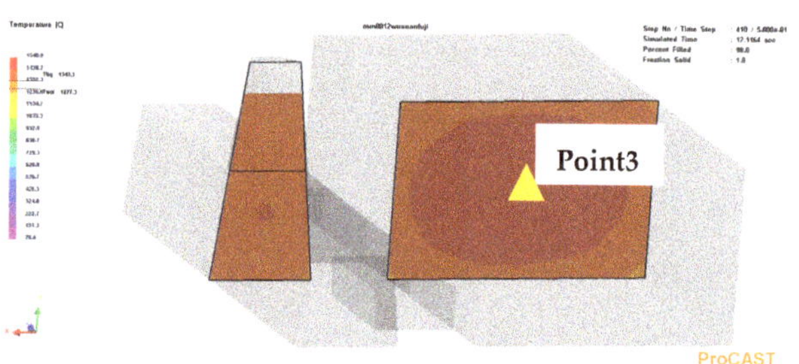


**Figure 11.** Schematic diagram of point taken of castings without adding suspension agent.

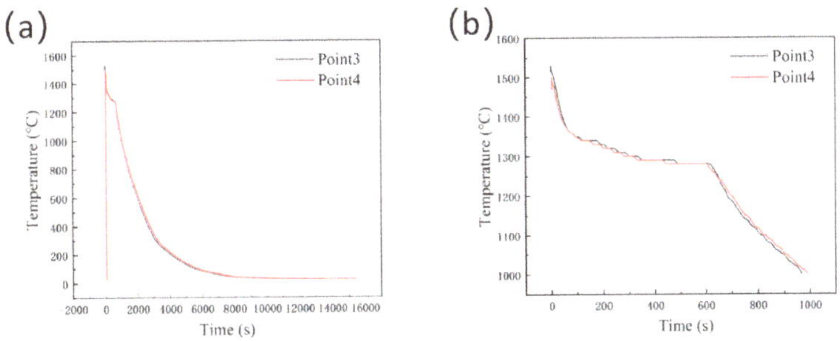


**Figure 12.** Temperature profiles of model without suspension and with 1% suspension agent added: (**a**) overall temperature results; (**b**) temperature profiles near high-temperature range.

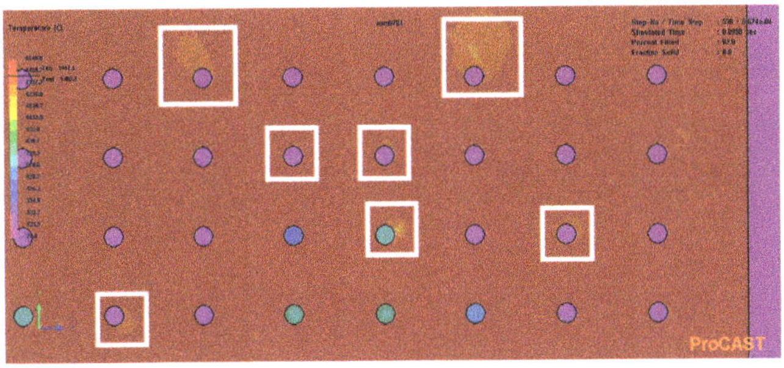


**Figure 13.** Temperature field near the suspension agent (Typical chilling effect can be observed in the white frame).

## 5. Conclusions

An innovative numerical simulation study on the lost foam suspension casting process of hypereutectic high-chromium cast iron was carried out. ProCAST software was used to observe the filling and solidification process of the simple cuboid ingot model, and the following important results were obtained.

(1) Through the simulation results of shrinkage porosity, it was found that the optimal process parameters within the study range were as follows: a negative pressure of 0.06 MPa, a pouring temperature of 1540 °C and 1% suspension agent.

(2) The addition of the suspension agent has an obstructive effect on the filling process of the lost foam casting, and the filling time is about 9% longer than that without the addition of the suspension agent.

(3) The pressure and thickness of the air gap layer remained stable during the process of filling with liquid metal. When the bottom casting system was adopted, the filling process was slow at first, then fast and then slow. The filling process was smooth and without turbulence or splashing phenomenon.

(4) The suspension agent melted during the solidification process. When the amount of TiFe was 1%, 89.4% of the suspension agent completely melted and the chilling effect of the suspension agent occurred within 0.44 s, resulting in a relative supercooling degree of about 50 °C for the liquid metal. It can be seen from the temperature curve that the solidification rate of hypereutectic high-chromium cast iron had increased under the lost foam suspension casting process. In order to further improve the chilling effect, it is necessary to increase the amount of suspension agent.

**Author Contributions:** Literature search, methodology, data collection and writing—original draft preparation, H.M.; writing—review and editing and project administration, H.F. All authors have read and agreed to the published version of the manuscript.

**Funding:** This research was funded by the National Natural Science Foundation of China, grant number 52075010.

**Data Availability Statement:** All data is contained within the article.

**Conflicts of Interest:** The authors declare no conflict of interest.

## References

1. Wang, X.H.; Jiao, Y.L. Effect of Ce Modification on Microstructure and Properties of Boron Containing Hypereutectic High Chromium Cast Iron. *Foundry* **2023**, *72*, 810–818.
2. Jian, A.S.; Chang, H.W.; Ahmad, H.; Ma, X.D.; Zhang, M.X. Effect of solutes on the formation of primary carbides during solidification of hypereutectic high chromium cast irons through thermodynamic modeling. *J. Mater. Sci.* **2022**, *57*, 1429–1447. [CrossRef]
3. Yilmaz, S.O.; Teker, T. Effect of TiBAl inoculation and heat treatment on microstructure and mechanical properties of hypereutectic high chromium white cast iron. *J. Alloys Compd.* **2016**, *672*, 324–331. [CrossRef]
4. Zhou, X.K.; Wang, Y.; Wu, S.; Xiao, X.L. Effect of Nb Content on Microstructure and Mechanical Properties of Hypereutectic High Chromium Cast Iron. *Hot Work. Technol.* **2022**, *51*, 50–52.
5. Li, P.; Yang, Y.H.; Shen, D.P.; Gong, M.Y.; Tian, C.; Tong, W.P. Mechanical behavior and microstructure of hypereutectic high chromium cast iron: The combined effects of tungsten, manganese and molybdenum additions. *J. Mater. Res. Technol.* **2020**, *9*, 5735–5748. [CrossRef]
6. Jain, A.S.; Mustafa, M.I.; Sazili, M.I.I.M.; Chang, H.W.; Zhang, M.X. Effects of destabilization and tempering on microstructure and mechanical properties of a hypereutectic high-chromium cast iron. *J. Mater. Sci.* **2022**, *57*, 15581–15597. [CrossRef]
7. Liu, Q.; Peter, H.; Zhang, H.W.; Wang, Q.; Pär, G.J.; Keiji, N. Effect of Heat Treatment on Microstructure and Mechanical Properties of Ti-alloyed Hypereutectic High Chromium Cast Iron. *ISIJ Int.* **2012**, *52*, 2288–2294. [CrossRef]
8. Deng, J.X.; Cai, H.T.; Tian, Y.L. Study on Casting Process of Hypereutectic High Chromium Cast Iron. *Foundry Eng.* **2022**, *46*, 34–37.
9. Zhang, K.; Zhang, J.P.; Zhang, Z.; Zhong, N.; Lu, S.W. Effect of Heat Treatment on Properties of Hypereutectic High Chromium Cast Iron. *Foundry* **2022**, *71*, 1133–1138.
10. Zhang, K.; Zhang, J.P.; Zhong, N.; Lu, S.W. Effect of quenching and tempering process on microstructure and properties of Cr26 hypereutectic high chromium cast iron. *Heat Treat. Met.* **2022**, *47*, 156–163.
11. Geng, B.Y.; Zhou, R.F.; Li, L.; Lv, H.Y.; Li, Y.K.; Bai, D.; Jiang, Y.H. Change in Primary (Cr, Fe)$_7$C$_3$ Carbides Induced by Electric Current Pulse Modification of Hypereutectic High Chromium Cast Iron Melt. *Materials* **2018**, *12*, 32. [CrossRef] [PubMed]
12. Geng, B.Y.; Zhou, R.F.; Li, Y.K.; Jiang, Y.H. Effect of electric current pulse on size of primary carbides in hypereutectic high chromium cast irons in different solidification stages. *Mater. Res. Express* **2019**, *6*, 126551. [CrossRef]
13. Gong, L.Q.; Fu, H.G.; Zhi, X.H. Corrosion Wear of Hypereutectic High Chromium Cast Iron: A Review. *Metals* **2023**, *13*, 308. [CrossRef]
14. Wang, J.H. Suspension casting process. *MW Met. Form.* **1980**, *05*, 60–62.
15. Fu, H.G.; Xie, P.M. Development and Application of Suspension Casting Process. *Mod. Cast Iron* **1966**, *01*, 33–34+46.
16. Jiang, Z.W.; Luo, H.F.; Li, Z.J. The Effect of Suspension Casting on the Properties of Fly-Ash Particle Reinforced Al Matrix Composites. *Adv. Mater. Res.* **2012**, *2044*, 581–582. [CrossRef]
17. Li, Q.S.; Liu, M.S. Effects of Suspension Casting on the Structure and Properties of High Chromium White Cast Iron. *J. Taiyuan Univ. Sci. Technol.* **2002**, *23*, 56–59.
18. Dai, B.Y. Application of Suspension Casting to Firebrick Forming Dies of High-Chromium Cast Iron. *Foundry* **2009**, *58*, 726–728.
19. Zhi, X.H.; Xing, J.D.; Fu, H.G.; Gao, Y.M. Effect of fluctuation, modification and surface chill on structure of 20%Cr hypereutectic white cast iron. *Mater. Sci. Technol.* **2009**, *25*, 56–60. [CrossRef]
20. Yan, X.N. Present situation and development trend of lost foam casting technology. *Sci. Technol. Ind. Parks* **2017**, *10*, 120.
21. Ma, G.B.; Tan, J.B.; Yu, Y.L.; Li, C.H. Study on Microstructure of Lost Foam Cast WC$_p$/Fe Composites. *Foundry* **2020**, *69*, 149–153.
22. Suyitno; Sutiyoko. Effect of Pouring Temperature and Casting Thickness on Fluidity, Porosity and Surface Roughness in Lost Foam Casting of Gray Cast Iron. *Procedia Eng.* **2012**, *50*, 88–94.
23. Xie, M.G.; Zhu, C.G.; Zhou, J.X. Effect on Solidification Structure of Hypoeutectic Grey Cast Iron by Vacuum Condition in Lost Foam Casting. In Proceedings of the 2015 6th International Conference on Manufacturing Science and Engineering, Qingdao, China, 17–19 July 2015; pp. 1496–1500.
24. Wu, G.H.; Luo, J.R.; Xie, M. Porosity of aluminum alloy in lost foam casting process. *Trans. Nonferrous Met. Soc. China* **2000**, *10*, 645–649.
25. Li, M.X.; Guo, Q.W.; Chen, L.W.; Li, L.M.; Hou, H.; Zhao, Y.H. Microstructure and properties of graphene nanoplatelets reinforced AZ91D matrix composites prepared by electromagnetic stirring casting. *J. Mater. Res. Technol.* **2022**, *21*, 4138–4150. [CrossRef]
26. Wang, J.; Pan, Z.X.; Wang, Y.F.; Wang, L.; Su, L.H.; Dominic, C.; Zhao, Y.H.; Li, H.J. Evolution of crystallographic orientation, precipitation, phase transformation and mechanical properties realized by enhancing deposition current for dual-wire arc additive manufactured Ni-rich NiTi alloy. *Addit. Manuf.* **2020**, *34*, 101240. [CrossRef]

27. Liu, G.S.; Wang, W.C.; Zhang, Z.M.; Liu, J.H.; Li, Q.L. Effect of Suspension Casting on the Property of Low-Chromium White Cast Iron Produced by the EPC Process. *Foundry* **2001**, *50*, 666–669.
28. Dai, H.Z.; Dai, H.M. Suspension Casting Process for Producing Wear Resistant High Manganese Steel. CN101880839A, 10 November 2010.
29. Fu, H.G.; Wu, Z.W.; Li, E.; Lei, Y.P.; Lin, J.; Zhang, J.G. Method of Manufacturing Castings by Suspension Casting. CN104084541B, 13 April 2016.
30. Edward, G.; Dariusz, K.; Andriy, B.; Andrzej, S. Evaluation of the Number of Primary Grains in Hypoeutectic Chromium Cast Iron with Different Wall Thickness Using the ProCAST Program. *Materials* **2023**, *16*, 3217.
31. Albuquerque, C.E.S.; Silva, P.C.S.; Grassi, E.N.D.; Araujo, C.J.D.; Delgado, J.M.P.Q.; Lima, A.G.B. Optimizing the Gating System for Rapid Investment Casting of Shape Memory Alloys: Computational Numerical Analysis for Defect Minimization in a Simple-Cubic Cell Structure. *Metals* **2023**, *13*, 1138. [CrossRef]
32. Yang, Q.; Zhu, X.T.; Wang, F.; Ma, D.X.; Wu, J.T. A Study of Sliver in C-Shaped Grain Selectors during Investment Casting of Single-Crystal Superalloy. *Metals* **2023**, *13*, 1102. [CrossRef]
33. Gong, Y.N.; Malik, A.; Wang, Y.W.; Feng, S.J.; Zhao, D.H. Numerical Simulation and Experimental Studies of Gas Pressure Infiltration Al-356/SiC Composites. *Metals* **2022**, *12*, 2150. [CrossRef]
34. He, T.Z.; Chen, Y.Y. Influence of Mold Design on Shrinkage Porosity of Ti-6Al-4V Alloy Ingots. *Metals* **2022**, *12*, 2122. [CrossRef]
35. Yan, J.W. Estabishment of Foam Pattern Database and Research on Numerical Simulation of Lost Foam Casting. Master's Thesis, Harbin University of Science and Technology, Harbin, China, 2014.
36. Feng, B. Numerical Simulation of Mould Filling and Solidification of EPC For Aluminum Alloys Cylinder Head. Master's Thesis, Hebei University of Science & Technology, Shijiazhuang, China, 2012.
37. Wang, J.H. *Suspension Casting*; National Defense Industry Press: Beijing, China, 1982; p. 56.
38. Huang, J.; Lin, Y.X.; Chen, W.P.; Qi, X.J. Numerical analysis of lost foam casting for large-caliber water meter shell. *Adv. Mech. Eng.* **2021**, *13*, 16878140211028059. [CrossRef]
39. Zhao, Z.W.; Jia, L.M.; Yu, Y.L.; Li, C.H.; Zhao, H.S.; Tan, J.B. Process design and numerical simulation of lost foam casting for grey cast iron end caps. *Hebei J. Ind. Sci. Technol.* **2022**, *39*, 341–348.
40. Hu, G.Q.; Peng, F.; Pan, F.Y.; Yang, Z.G.; Zhu, W.W. Influence of Negative Pressure Degree on Quality of Vacuum Negative Pressure Lost Foam Gray Iron Casting. *Mod. Cast Iron* **2018**, *38*, 33–37.
41. Deng, C.; Long, J.; Zheng, Z.B.; Huang, Y.; Zheng, K.H. Numerical Simulation of Solidification Process in Lost Foam Casting of Caterpillar Board Based on ProCAST Software. *Spec. Cast. Nonferrous Alloys* **2021**, *41*, 368–371.
42. Liang, H.; Li, Z.M.; Wang, Y.C.; Zahng, X.M.; Li, L.X.; Tan, J.B.; Xue, X.X. Filling Morphology of Gray Cast Iron Melt in Lost Foam Casting. *Spec. Cast. Nonferrous Alloys* **2007**, *27*, 870–871+818.
43. Chang, Q.M.; Huang, N.Y.; Luo, J.R.; Ye, S.P. Numerical Simultion of Mold Filling in EPC Process. *Foundry* **1997**, *2*, 7–11.
44. Li, F.J.; Zhao, H.; Ren, F.Z.; Song, S.B.; Shao, X.H. Simulations and experiments of mould filling in lost foam casting. *Int. J. Cast Met. Res.* **2020**, *33*, 194–200. [CrossRef]
45. Dai, G.H. Application of Lost Foam Casting Process for Ductile Iron Shell Based on ProCAST. Master's Thesis, Anhui University of Technology, Maanshan, China, 2018.

*metals*

MDPI

*Article*

# Microstructure and Anisotropy of Mechanical Properties of Al-3Li-1Cu-0.4Mg-0.1Er-0.1Zr Alloys Prepared by Normal Rolling and Cross-Rolling

**Zhikun Ma [1], Tao Zhong [2], Dongpeng Sun [1], Bingyu Qian [1], Nodir Turakhodjaev [3], Sergey Betsofen [4] and Ruizhi Wu [1,*]**

1 Key Laboratory of Superlight Materials & Surface Technology, Ministry of Education, Harbin Engineering University, Harbin 150001, China; zkma@hrbeu.edu.cn (Z.M.)
2 Department of Science and Technology, Nanning College for Vocational Technology, Nanning 530008, China
3 Department of Machine Building, Tashkent State Technical University, Tashkent 100056, Uzbekistan
4 Moscow Aviation Institute, Russian National Research University, Moscow 121552, Russia
* Correspondence: rzwu@hrbeu.edu.cn

**Abstract:** The influence of normal rolling and cross-rolling on the microstructure, mechanical properties, and anisotropy of Al-3Li-1Cu-0.4Mg-0.1Er-0.1Zr alloy was investigated. With an increase in the rolling reduction amount, both the strength and plasticity of the alloy are enhanced. Among them, the alloy in the normal rolling state with a deformation amount of 90% exhibits the best properties, with a tensile strength of 362 MPa and an elongation of 19.1% along the rolling direction. During the rolling process, the intergranular Cu-containing phase in the alloy is continuously broken and dissolved, leading to a decrease in both size and quantity, turning from continuous distribution along grain boundaries to a granular distribution. Moreover, a large quantity of the $Al_3Li$ phase and $Al_3$(Er, Zr, Li) core–shell composite phase are precipitated in the alloy. Recrystallization occurs mainly through the particle stimulated nucleation (PSN) mechanism. Cross-rolling eliminates the brass-type texture <111> produced by normal rolling and enhances the brass R-type texture {111}<112>. The index of plane anisotropy (IPA) of the strength decreases from 10.1% for normal rolling to 5.5% for cross-rolling, and the IPA of elongation decreases from 12.8% to 3.3%. Cross-rolling provides an effective method to reduce the anisotropy of Al-Li alloys.

**Keywords:** Al-Li alloy; cross-rolling; microstructure; anisotropy; texture

## 1. Introduction

Al-Li alloys are considered to be the ideal lightweight, high-strength structural materials in the aerospace field due to their high elastic modulus, high specific strength, low density, good weldability, and corrosion resistance [1–3]. After deformation processing, the microstructure of Al-Li alloys can be improved to enhance their mechanical properties and ultimately obtain the desired sheets or plates [4]. The main deformation processes for Al-Li alloys include rolling, extrusion, and stretching. Among them, rolling is one of the most commonly used deformation processes [5–7].

During the hot rolling process, the microstructure and mechanical properties undergo a series of changes, such as work hardening, dynamic softening, dissolution and precipitation of second phases, and grain preferential orientation [8–10]. An appropriate hot rolling process can not only reduce the anisotropy but also enhance mechanical properties of the alloy. Ye et al. [11] investigated the effects of different methods of thermomechanical processing on the precipitation behavior, crystallographic texture, and tensile properties of Al-Cu-Li alloy (AA2195). The results showed that although HC-T8-processed alloy had higher tensile strength and elongation compared to H-T8-processed alloy, the strength anisotropy of HC-T8 alloy was more pronounced. The higher dislocation density existed in the H-T8 samples, which promoted the nucleation of strengthening phase T1 in the matrix and suppressed the precipitation at grain boundary, resulting in better strength and ductility. Zhuo et al. [12] investigated the effects of the combined processing methods of heat treatment and rolling on the mechanical properties of Al-6Si-3Cu alloy. After solution treatment and rolling deformation, the as-cast Al alloy exhibited a tensile strength increased by 105%, reaching 429.8 MPa, and an improved elongation by 76%, reaching 5.8%. The increased ductility compared to as-cast alloy is due to the fragmentation and refinement of Si particles and dissolution of coarse precipitates.

Different rolling deformation processes have different effects on the texture of alloys, thus influencing the anisotropy of alloys [13–15]. Huh et al. [16] found that an AA5182 alloy sheet improved its strong texture through cross-rolling accompanied by recrystallization texture. Niraj et al. [17] found that AA2195 alloy exhibited typical copper texture after conventional rolling, while it showed a brass texture after cross-rolling. Li et al. [18] investigated the effects of cross-rolling and annealing on the microstructure and mechanical properties of AA7075 alloy and found that cross-rolling reduced the intensity of dominant textures, such as brass and S grain orientations, causing the decreased anisotropy of the alloy. However, texture is not the sole factor affecting the anisotropy of Al-Li alloys. Other factors, such as the presence of grain-boundary-precipitation-free zones and the type and size of second phases, also influence the anisotropy [19–21].

In this work, Al-3Li-1Cu-0.4Mg-0.1Er-0.1Zr alloy was processed with normal rolling and cross-rolling at 500 °C. The alloy with a low anisotropy was obtained through the cross-rolling process. The evolution of the microstructure, mechanical properties, and anisotropy under different rolling processes was investigated.

## 2. Materials and Methods

Al-3Li-1Cu-0.4Mg-0.1Er-0.1Zr alloy was melted and cast using commercial pure ingots of Al (99.9 wt.%), Li (99.9 wt.%), and Mg (99.9 wt.%) and master alloy ingots of Al-50 wt.%Cu, Al-10 wt.%Er, and Al-10 wt.%Zr. Subsequently, the as-cast alloy was subjected to homogenization annealing at 500 °C for 24 h followed by furnace cooling. The chemical composition of the alloy was determined using inductively coupled plasma atomic emission spectroscopy, as shown in Table 1.

**Table 1.** The chemical composition of the as-received Al-3Li-1Cu-0.4Mg-0.1Er-0.1Zr alloy.

| Elements | Li | Cu | Mg | Er | Zr | Al |
|---|---|---|---|---|---|---|
| Content (wt.%) | 3.198 | 1.165 | 0.4087 | 0.1188 | 0.1237 | Bal. |

In the normal rolling, the rolling direction was kept unchanged. In this study, sheet materials with initial thicknesses of 2.5 mm, 4 mm, and 20 mm were rolled to a final thickness of 2 mm, resulting in corresponding deformation rates of 20%, 50%, and 90%, respectively. Each reduction in thickness for each pass was set at 0.4 mm. The different initial thicknesses of the sheet materials resulted in deformation rates of 16%, 10%, and 2% for the respective thicknesses. The rolling was conducted at 500 °C with a roll diameter of 500 mm and a roll speed of 600 r/min. Prior to rolling, the specimens were heated to 500 °C using a resistance furnace, and the temperature was held for 30 min. At the intervals between two passes, the specimens were kept in the resistance furnace for 5 min. The final rolled sheet thickness was 2 mm.

In the cross-rolling, the initial thickness was set as 20 mm. Other parameters were the same as in the normal rolling process except for the change in rolling direction between passes. Two rolling directions were selected: one was the initial rolling direction (RD), and the other was perpendicular to the previous rolling direction (TD). The rolling process was performed alternately between these two directions, and the alloy was ultimately rolled to a thickness of 2 mm.

The mechanical properties of the specimens were measured by a WDW3050 electronic universal testing machine. The tensile specimen dimensions are shown in Figure 1. The nominal strain rate was $1 \times 10^{-3}s^{-1}$. Five parallel samples were selected for testing in the same state, and the average value was taken. In order to investigate the effect of rolling processes on the anisotropy of Al-Li alloy, the conventional rolled alloy with 90% deformation and the cross-rolled alloy were taken for tensile testing. The specimens were cut at 0°, 45°, and 90° with respect to the rolling direction (RD direction), as shown in Figure 1.

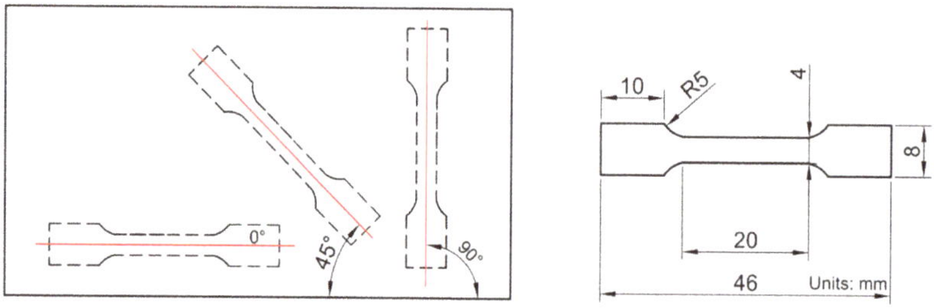


**Figure 1.** The locations and dimensions of the tensile specimens.

Phase analysis was carried out by X-ray diffraction (XRD, Rigaku TTR-III) with a copper foil target, tube voltage 40 kv, tube current 110 mA, scanning speed of 2°/min, start angle of 10°, and end angle of 90°. The microstructure and tensile fracture morphology of the rolled alloy were observed using a Hitachi SU5000 scanning electron microscope (SEM) with an attached energy dispersive spectrometer (EDS) to measure the chemical compositions of the phases. The morphology, composition, and distribution of the second phase in the rolled alloy were observed and characterized using a JEM-2100EX transmission electron microscope (TEM). The texture measurements of the specimens were conducted using a Hitachi SU5000 SEM electron backscatter diffraction (EBSD) system. After the testing, HKL Channel 5 software was used to process and analyze the collected data, outputting inverse pole figures, pole figures, and orientation distribution function (ODF) maps.

## 3. Results and Discussion

### 3.1. *Microstructure and Mechanical Properties of Different Rolling Reductions*

#### 3.1.1. Microstructural Evolution during Rolling with Different Reductions

As shown in Figure 2, the gray-white phase represents Cu-rich phases, while the bright-white phase is a mixed phase mainly composed of Cu-rich phases. Figure 2a,b show that after 20% deformation rolling, a small amount of Cu-rich phases near the grain boundaries are fragmented in blocklike shapes, while the majority of the Cu-rich phases still maintain continuous distribution along the grain boundaries. After 50% deformation rolling, more Cu-rich phases that were originally continuously distributed along the grain boundaries are fragmented in particle-like shapes, as shown in Figure 2c,d. It can also be observed that the Cu-rich phases that have not been completely fragmented still distribute along the grain boundaries, but they have transformed from large blocklike shape to fine elongated strips. Overall, there is a reduction in the quantity of Cu-rich phases. After 90% deformation rolling, the Cu-rich phases that are originally continuously distributed along the grain boundaries in the Al-Li alloy are completely fragmented into particles, as shown in Figure 2e,f. From general observation, the quantity of Cu-rich phases in the alloy after 90% deformation rolling is significantly reduced, and there are almost no larger Cu-rich phases.

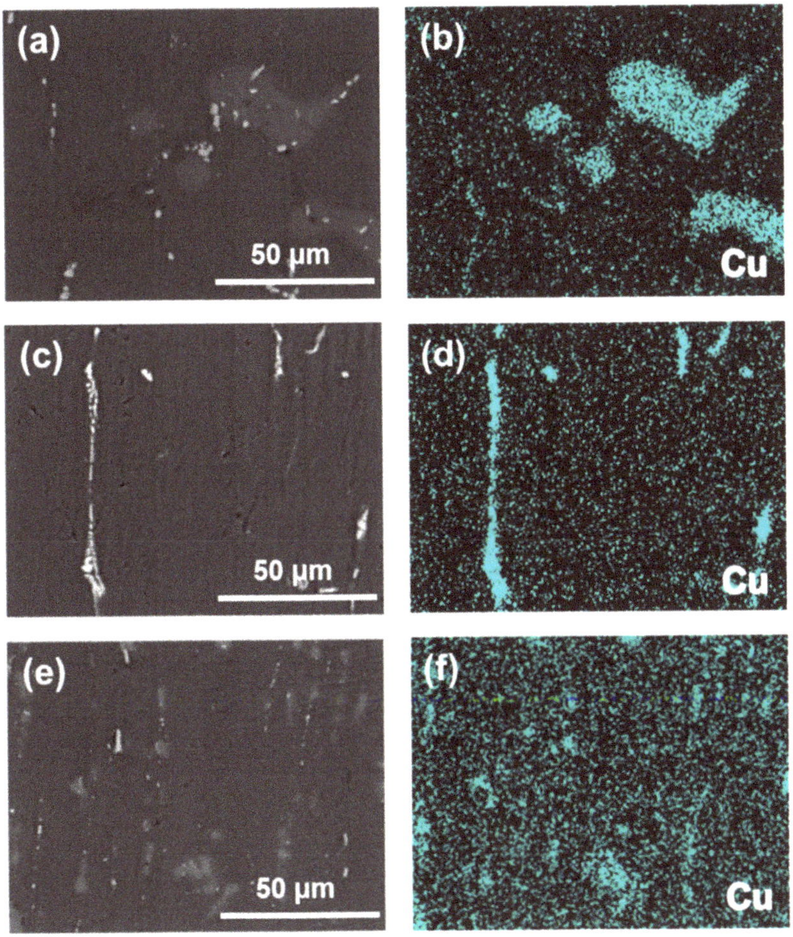


**Figure 2.** SEM images and their corresponding EDS mapping of Cu element in the RD–ND plane of Al-3Li-1Cu-0.4Mg-0.1Er-0.1Zr alloys with different rolling deformations: (**a**,**b**) 20%, (**c**,**d**) 50%, and (**e**,**f**) 90%.

XRD patterns of Al-Li alloys with deformation amounts of 20%, 50%, and 90% are presented in Figure 3. In 20% and 50% deformed Al-Li alloys, only the diffraction peaks of α-Al and $Al_2Cu$ are observed. By comparing the XRD patterns of the 20%, 50%, and 90% deformed Al-Li alloys, it can be noticed that the intensity of the $Al_2Cu$ phase diffraction peak significantly decreases in the 90% deformed Al-Li alloy, indicating that a large amount of $Al_2Cu$ phase is dissolved into the matrix. Moreover, besides the diffraction peaks of α-Al and $Al_2Cu$, some new diffraction peaks of $Al_3Li$ phase appear, indicating that a large amount of the $Al_3Li$ phase is dynamically precipitated. However, according to the literature, there are also $Al_3Zr$ phase, $Al_3Er$ phase, $Al_3$(Er, Zr) phase, and $Al_3$(Er, Zr, Li) composite phase with the same crystal structure as the $Al_3Li$ phase, which all have an $L1_2$ crystal structure [5]. Therefore, based on the XRD patterns of the 90% deformed Al-Li alloy in the rolled state, it is possible that besides the dynamic precipitation of the $Al_3Li$ phase, the dynamic precipitations of the $Al_3Zr$ phase, $Al_3Er$ phase, $Al_3$(Er, Zr) phase, and $Al_3$(Er, Zr, Li) composite phase also occur during the hot rolling process.

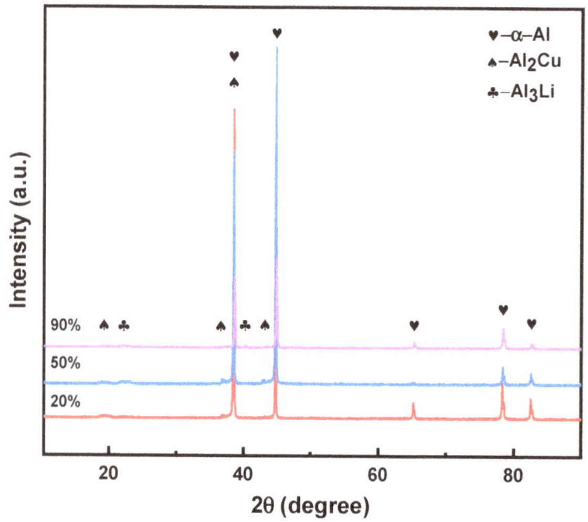


**Figure 3.** The XRD patterns of Al-3Li-1Cu-0.4Mg-0.1Er-0.1Zr alloys with rolling deformation amounts of 20%, 50%, and 90%, respectively.

Figure 4 shows the TEM dark-field image of the Al-Li alloys in the as-rolled state at 90% deformation and the selected-area electron diffraction (SAED) pattern along the $[011]_{Al}$ axis. It can be observed that there is a large number of bright spherical nanoscale second phases, which exhibit the typical morphology of the $Al_3Li$ phase. From Figure 4b, the electron diffraction pattern can be identified as two sets of diffraction spots: one set originates from the face center cubic (FCC) lattice of the matrix, which is α-Al, and the other set represents the $L1_2$-type superlattice diffraction spots embedded in the FCC diffraction spots. It can be concluded that the dispersed bright spherical second phase in Figure 4a is an $Al_3Li$ phase with an $L1_2$ crystal structure. Additionally, there are also some dark nanoscale second phases, which are much larger in size compared to the $Al_3Li$ phase. These phases have a dark center and a white circular shell, showing a typical core–shell structure. These phases are possibly $Al_3$(Er, Li) composite phases, $Al_3$(Zr, Li) composite phases, or $Al_3$(Er, Zr, Li) composite phases, where the inner layer consists of $Al_3Er$, $Al_3Zr$, or $Al_3$(Er, Zr) composite phases and the outer white shell consists of an $Al_3Li$ phase [22,23].

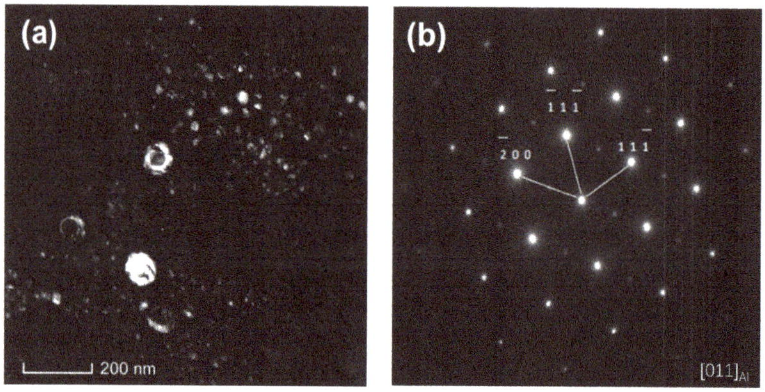


**Figure 4.** The dark-field TEM images of a 90% rolled state Al-3Li-1Cu-0.4Mg-0.1Er-0.1Zr alloy and the electron diffraction spots of the $[011]_{Al}$ axis: (**a**) dark phase, and (**b**) electron diffraction pattern.

As shown in Figure 5, the high-resolution transmission electron microscope (HRTEM) image reveals the presence of a core–shell composite phase, and corresponding diffraction patterns were obtained by performing a Fourier transform at different positions within the core–shell structure. Further analysis was conducted to investigate the structural composition of the core–shell phase. In Figure 5a, a clear observation of the core–shell structured phase can be made, where the core region and shell region of the core–shell structured phase are marked with dashed lines. It can be clearly seen that the core region of the $Al_3(Er, Zr)$ composite phase has a darker contrast, followed by the brighter shell region, and then the darker α-Al matrix region. Subsequently, a Fourier transform was performed on these three regions, resulting in the respective diffraction patterns. It can be observed from Figure 5d that the Fourier transform diffraction at the α-Al matrix region surrounding the core–shell structured phase is a typical FCC diffraction. From Figure 5b,c, it can be observed that the Fourier transform diffraction patterns at the core position and shell position of the core–shell structured phase correspond to $L1_2$-type superlattice diffractions. Based on the above analysis, although both the core and shell of the core–shell structure exhibit $L1_2$-type superlattice diffraction, the diffraction pattern obtained through the Fourier transform indicates that the core structure corresponds to the $Al_3Li$ phase while the shell corresponds to the $Al_3(Er, Zr)$ phase. Therefore, the core–shell structure can be described as a composite phase of $Al_3(Er, Zr, Li)$ [24–26].

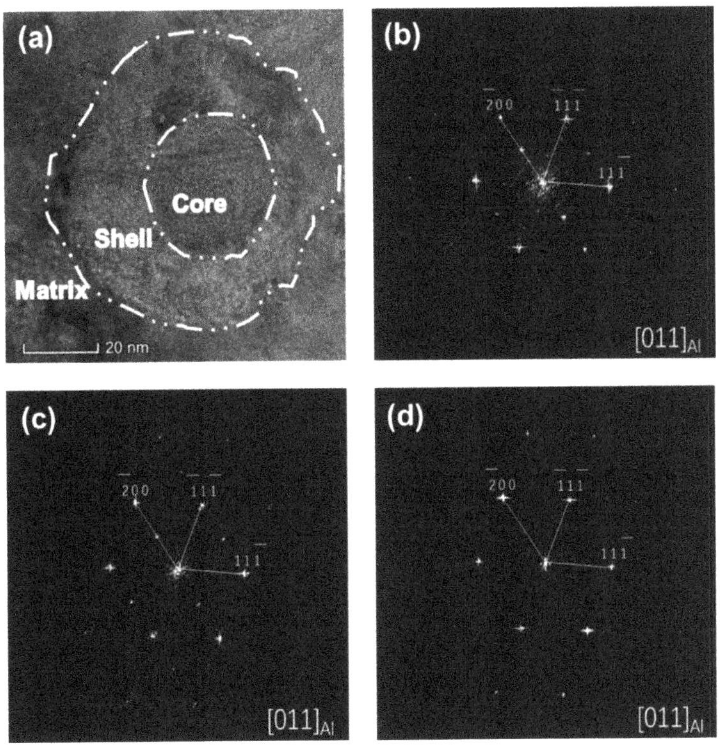


**Figure 5.** HRTEM images of core–shell structured composite phases in 90% cold rolled Al-3Li-1Cu-0.4Mg-0.1Er-0.1Zr alloy and the corresponding Fourier transform diffraction patterns obtained at different positions of the core–shell structure: (**a**) HRTEM image, (**b**) Fourier transform diffraction pattern at the core position, (**c**) Fourier transform diffraction pattern at the shell position, and (**d**) Fourier transform diffraction pattern at the matrix position.

3.1.2. Mechanical Properties during Rolling with Different Reductions

The stress–strain curves of Al-3Li-1Cu-0.4Mg-0.1Er-0.1Zr alloys with different rolling deformations are shown in Figure 6. With the increase in rolling deformation, the mechanical properties of the alloys, including tensile strength, yield strength, and elongation, continuously improve, with the best mechanical properties achieved at the 90% deformation. The as-rolled Al-Li alloy with 20% deformation has a tensile strength of 102 MPa and an elongation of 1.1%, but it does not exhibit yielding. When the rolling deformation increases to 50%, the tensile strength, yield strength, and elongation are 194 MPa, 92 MPa, and 5.4%, respectively. When the rolling deformation increases to 90%, the mechanical properties of the alloy show significant improvement. The tensile strength, yield strength, and elongation are 362 MPa, 274 MPa, and 19.1%, respectively. With the increase in rolling deformation, the mechanical properties of the alloy, including tensile strength, yield strength, and elongation, are greatly enhanced.

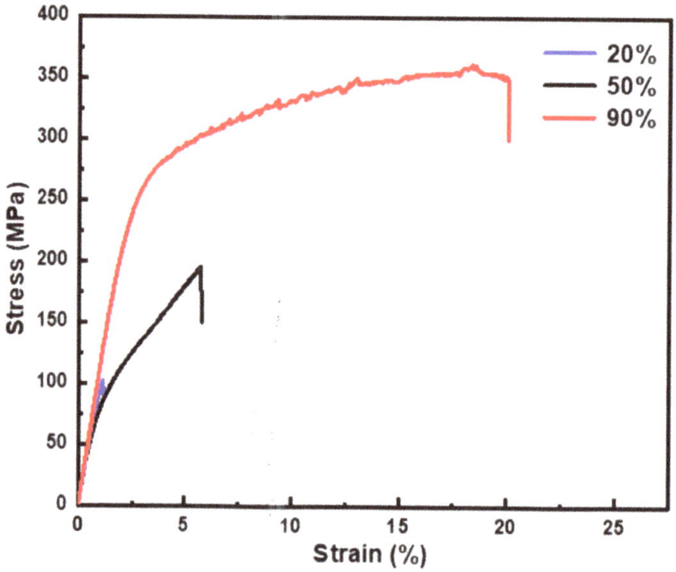


**Figure 6.** Stress–strain curves of Al-3Li-1Cu-0.4Mg-0.1Er-0.1Zr alloys under different rolling deformation percentages.

Figure 7a,b display the tensile fracture surfaces of Al-Li alloy with a 20% rolling deformation. In the images, typical low-melting-point metal brittle intergranular fractures can be observed, which are caused by the segregation of low-melting-point eutectic phases at the grain boundaries. The fracture mode of the Al-Li alloy with a 20% rolling deformation is the intergranular brittle fracture. Additionally, large blocks of the second phase along the grain boundaries can also be observed in the tensile fracture surfaces, which greatly influence the mechanical properties of the alloy, particularly its plasticity. Figure 7c,d represent the tensile fracture surfaces of Al-Li alloy with a 50% rolling deformation, where many smooth intergranular fracture surfaces, coarse second-phase particles, and pores resulting from the detachment of the second phase can clearly be seen. It can be further observed that compared to the 20% rolling deformation, the fracture surface of the 50% rolling deformation shows more dimples, although their distribution is not uniform. Hence, it can be inferred that the fracture mode of the Al-Li alloy with a 50% rolling deformation is a mixture of intergranular and dimple fractures. Figure 7e,f illustrate the tensile fracture surfaces of Al-Li alloy with a 90% rolling deformation, where numerous dimples are observed with a more uniform distribution. Some intergranular fracture surfaces and traces of fractured and detached second-phase particles with sizes around 10 μm can also be observed. Compared to the 50% rolling deformation, the number and distribution of dimples increase in the tensile fracture surface of the Al-Li alloy with a 90% rolling deformation. The size and quantity of second-phase particles decrease. Based on these results, it can be preliminarily concluded that the fracture mode of the Al-Li alloy after a 90% rolling deformation is a mixture of intergranular and dimple fractures, with the dimple fractures being predominant.

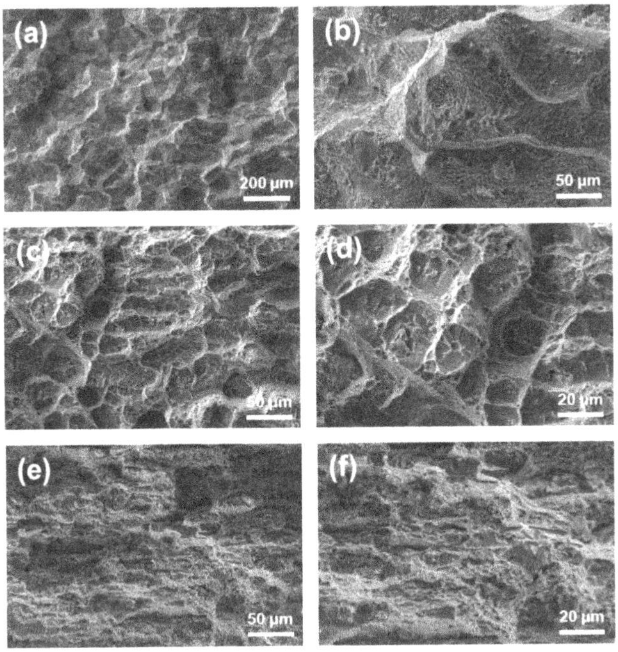


**Figure 7.** Scanning electron microscopy of tensile fracture surfaces of aluminum–lithium alloys at different rolling deformation levels: (**a**,**b**) 20%, (**c**,**d**) 50%, and (**e**,**f**) 90%.

### *3.2. Microstructure and Mechanical Properties of Normal Rolled and Cross-Rolled Specimens*

#### 3.2.1. Microstructure of Normal Rolled and Cross-Rolled Specimens

Figure 8 shows the EBSD grain maps of normal rolled and cross-rolled Al-Li alloys. In Figure 8a, it can be observed that the grains in the normal rolled alloy are severely elongated along the RD direction, presenting a typical fibrous morphology. Additionally, numerous small equiaxed grains can be observed, indicating recrystallization occurrence. In Figure 8b, it can be seen that the elongation of the grains along the RD direction is relatively not obvious in the cross-rolled alloy, which is attributed to the increased deformation component in the TD direction. A few small recrystallized grains are also observed. A comparison reveals that the grain orientation in the normal rolled Al-Li alloy is mainly oriented along the <111>//RD direction, with a small fraction oriented along the <001>//RD direction. However, the grain orientation in the cross-rolled Al-Li alloy exhibits a more pronounced diversification, with a noticeable decrease in the <111>//RD orientation and a shift towards other directions. This indicates that normal rolling tends to align the grains in the Al-Li alloy towards a single orientation, while the cross-rolling process can alter this orientation pattern, leading to a more diversified grain orientation in the rolled alloy [27–29].

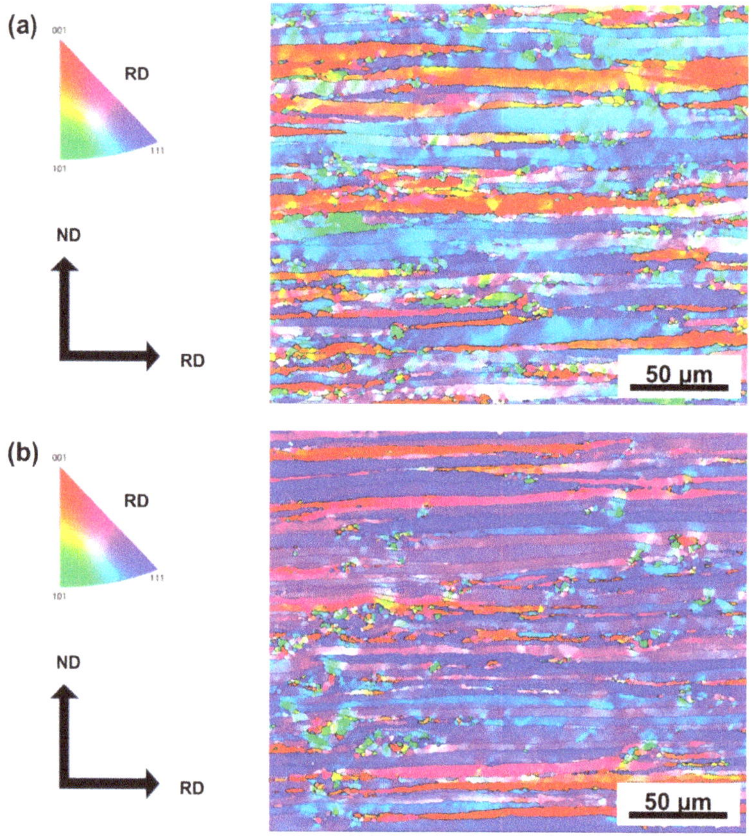


**Figure 8.** EBSD grain map of 90% deformation Al-3Li-1Cu-0.4Mg-0.1Er-0.1Zr alloy in conventional rolled and cross-rolled state: (**a**) normal rolling, and (**b**) cross-rolling.

Figure 9 shows the recrystallization of Al-Li alloys with 90% deformation in both normal rolling and cross-rolling. Although recrystallization occurs after cross-rolling, the degree of recrystallization is significantly reduced compared to normal rolling, with a noticeable decrease in proportion of recrystallized grains, 23.5% vs. 4.5%. Figure 10 shows the EBSD microstructure at the same position. Some voids can be observed in Figure 10, which are caused by the detachment of coarse second-phase particles during electropolishing. By combining Figures 9 and 10, it can be observed that the regions with detached second-phase particles in Figure 10 correspond well to the regions with obvious recrystallized grains in Figure 9. This indicates that recrystallization is more likely to occur near the coarse second-phase particles during rolling, known as particle stimulated nucleation (PSN) [30]. The recrystallized grains in the cross-rolled alloys are also distributed near micrometer-sized second-phase particles in the alloy, indicating that the recrystallization mechanism is also dependent on PSN. However, by comparing the observations, it can also be found that near some second-phase particles in the cross-rolled Al-Li alloy, there do not exist recrystallized grains, indicating that these particles do not induce recrystallization via PSN. Compared to conventional rolling, the degree of recrystallization in the cross-rolled Al-Li alloy is reduced, with a decrease in the proportion of recrystallized grains by approximately 6.5%. It can be concluded that, in the cross-rolling process, the alloy undergoes alternating deformation along the RD–ND directions, with different rolling directions between adjacent passes, which activates more slip systems compared to normal rolling, thereby promoting recovery and reducing the dislocation density in the alloy. Although the PSN mechanism can help accumulate deformation energy and promote recrystallization, the overall reduction in dislocation density in the alloy makes it difficult to reach the critical value for recrystallization. Therefore, only a portion of the second-phase particles in the cross-rolled alloy induce recrystallization, resulting in a reduced number of recrystallized grains compared to normal rolling [31–33].

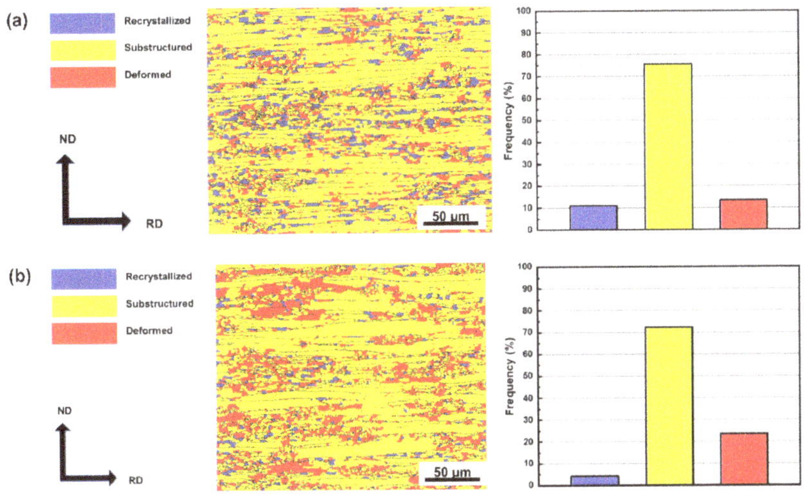


**Figure 9.** The recrystallization characteristics of 90% deformation in both normal rolling and cross-rolling states of Al-3Li-1Cu-0.4Mg-0.1Er-0.1Zr alloys: (**a**) normal rolling, and (**b**) cross-rolling.

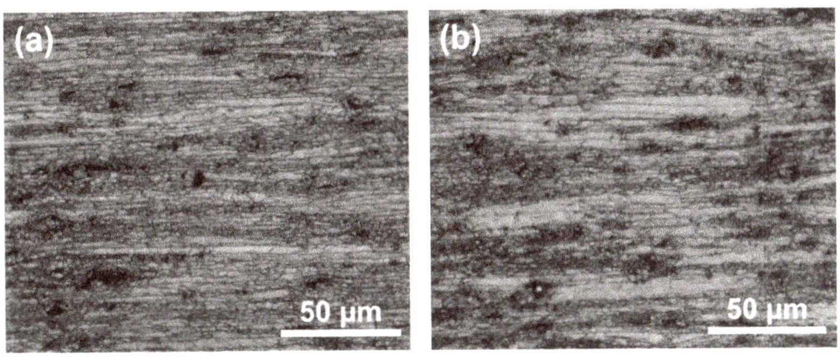


**Figure 10.** Microstructure of Al-3Li-1Cu-0.4Mg-0.1Er-0.1Zr alloys with 90% deformation by normal rolling and cross-rolling: (**a**) normal rolling, and (**b**) cross-rolling.

### 3.2.2. Grain Orientation and Texture of Al-Li Alloys Processed by Normal Rolling and Cross-Rolling

As shown in Figure 11, it can be observed that the phase composition of the alloys does not change in normal rolled and cross-rolled specimens, mainly consisting of α-Al, $Al_3Li$, and Cu-rich phases. However, we can observe that the intensity of diffraction peaks varies. After cross-rolling, the intensity of the (111) and (200) diffraction peaks is enhanced, while the intensity of the (200) diffraction peak is decreased, suggesting that the preferred grain orientation changes.

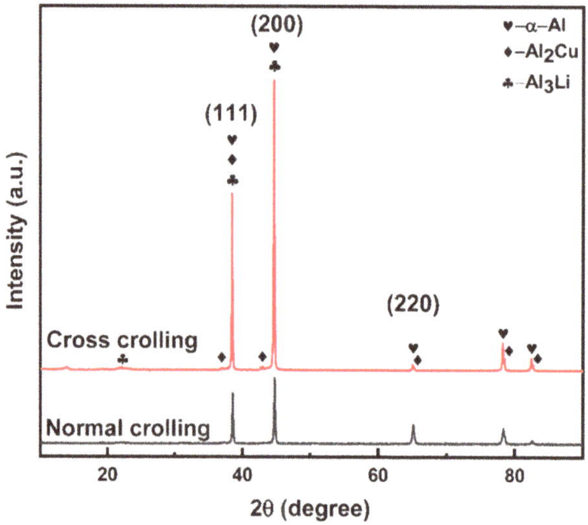


**Figure 11.** XRD pattern of conventional rolling and cross-rolling with 90% deformation.

The pole figures of the as-rolled and cross-rolled Al-Li alloys are presented in Figure 12. X0 and Y0 represent the rolling direction and normal direction, respectively. Analysis reveals that the normal rolled Al-Li alloy predominantly exhibits a brasslike texture with the {011}<111> orientation, while the cross-rolled alloy primarily exhibits a brass R-type texture with the {111}<112> orientation [30].

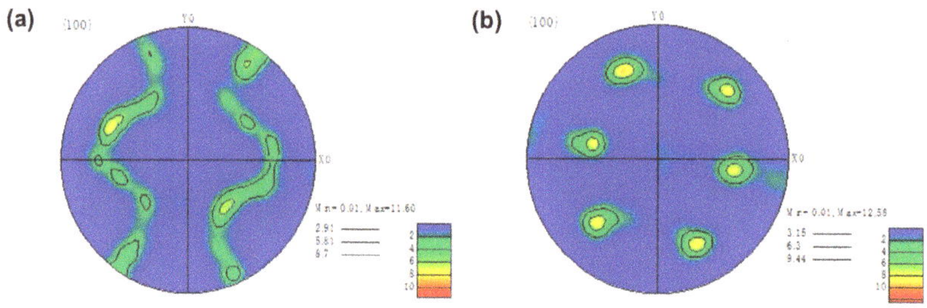


**Figure 12.** Different rolling processes for Al-3Li-1Cu-0.4Mg-0.1Er-0.1Zr alloy {001} surface pole: (**a**) normal rolling, (**b**) cross-rolling.

The ODF (orientation distribution function) graphs of the normal rolled and cross-rolled Al-Li alloys are shown in Figure 13. The graphs utilize a constant φ2 in a range of 0° to 90°, with a 5-degree interval per image. It is observed that the alloy exhibits a strong brasslike texture {011}<111> after normal rolling, as depicted in Figure 13a. This is consistent with the results shown in the pole figure of Figure 13a, with the highest intensity of 9.11. Additionally, a weaker brass R-type texture {111}<112> is present in the alloy after normal rolling. Figure 13b demonstrates that the alloy displays a strong brass R-type texture {111}<112> after cross-rolling, which aligns with the pole figure result in Figure 12b. The highest intensity in this texture reaches 10.07. Moreover, a weak cube texture {001}<100> can be observed in the cross-rolled alloy [34,35].

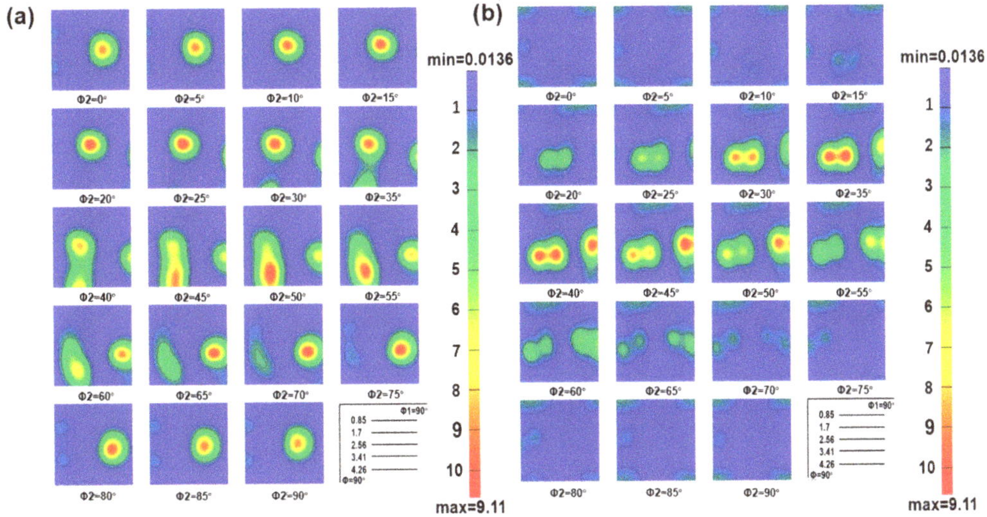

Figure 13. ODF diagrams of Al-3Li-1Cu-0.4Mg-0.1Er-0.1Zr alloys in different rolling processes: (a) normal rolling, and (b) cross-rolling.

Cross-rolling changes the alloy texture composition of alloy. After cross-rolling, the intensity of the brass R-type texture {111}<112> become stronger, eliminating the brass-type texture {011}<111>, while also adding a weak cubic cube texture {001}<100> [36]. The main texture of the alloy changed from the brass-type texture {011}<111> to the brass R-type texture {111}<112>, with similar intensity. According to the literature, the brass-type texture {011}<111> is a recrystallization texture. It is formed by rotating the brass-type texture around the ND direction and usually appears after recrystallization through the PSN mechanism [30]. Based on the characterization of recrystallization mentioned earlier, it can be concluded that normal rolling of Al-Li alloy contains a large number of recrystallized grains formed by the PSN mechanism, which coincides with the recrystallization texture of normal rolling. Although recrystallization also occurs during cross-rolling, it does not exhibit a recrystallization texture, indicating that the recrystallized grains generated by the cross-rolling process are randomly oriented.

### 3.2.3. Mechanical Properties of Normal Rolled and Cross-Rolled Specimens

Figure 14 shows the stress–strain curves of the normal rolled and cross-rolled Al-Li alloys, with three different orientations of 0° (RD), 45°, and 90° (TD) with respect to the RD direction. It can be observed that the tensile strength varies significantly in different directions for the normal rolled alloy, while the cross-rolled Al-Li alloys exhibit smaller variations in tensile strength among the different orientations.

To provide a more comprehensive understanding, the tensile strength and elongation of Al-Li alloys were statistically compared, as shown in Figure 15. The specimens along RD in the normal rolled state have the highest tensile strength, reaching 362 MPa, while those along TD have the lowest tensile strength at only 306 MPa. Comparatively, the tensile strength values in different directions of the cross-rolled Al-Li alloys are relatively close, with the RD still having the highest tensile strength of 342 MPa, while the lowest tensile strength of 319 MPa is observed in the 45° direction. From Figure 15b, it can be observed that the normal rolled specimen along RD has the highest elongation, reaching 19.1%, while that along the 45° direction has the lowest elongation of 14.7%. The cross-rolled specimen along the 45° direction has the highest elongation, reaching 15.2%, and that along TD has the lowest elongation of 14.6%.

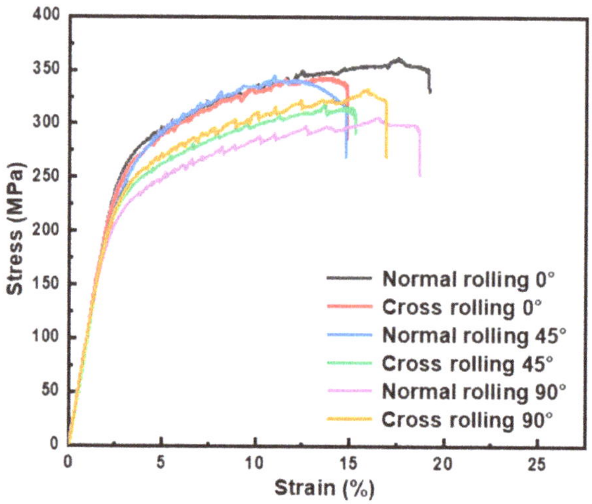


**Figure 14.** A 90% deformation stress–strain curve of normal rolled and cross-rolled Al-3Li-1Cu-0.4Mg-0.1Er-0.1Zr alloys.

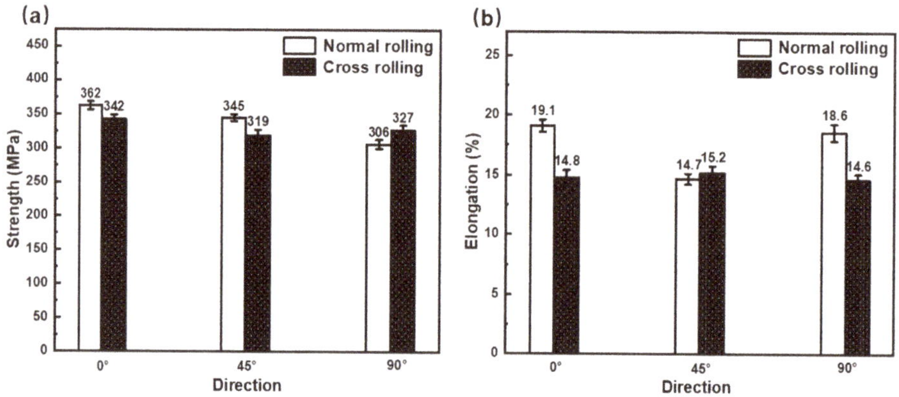


**Figure 15.** Tensile strength and elongation of Al-3Li-1Cu-0.4Mg-0.1Er-0.1Zr alloys in different rolling processes: (**a**) strength, and (**b**) elongation.

The planar anisotropy of the mechanical properties of alloy sheets can be measured using the IPA. The greater the IPA value, the greater the anisotropy of the mechanical properties [4]. The calculation method is shown in Equation (1).

$$\text{IPA} = \frac{2\text{Xmax} - \text{Xmid} - \text{Xmin}}{2\text{Xmax}} \times 100\% \quad (1)$$

where Xmax, Xmid, and Xmin represent the maximum, intermediate, and minimum values, respectively. Combined with Figure 15, the IPA values of tensile strength and elongation of the alloy can be calculated separately. Figure 16 shows the statistical chart of IPA values for the strength and elongation of the normal rolled and cross-rolled specimens. Compared with the strength IPA of normal rolled specimens, that of cross-rolled specimens decreases from 10.1% to 5.5%, indicating a significant decrease in its anisotropy. The IPA of elongation also decreases from 12.8% to 3.3%, further demonstrating the decreased anisotropy of Al-Li alloy after cross-rolling [36].

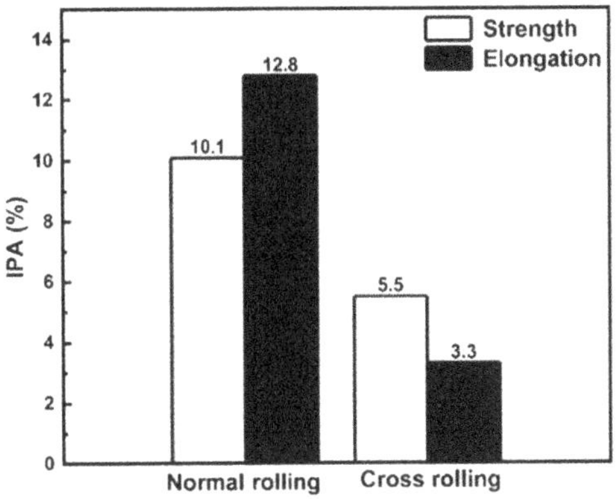


**Figure 16.** Statistical diagram of IPA values of Al-3Li-1Cu-0.4Mg-0.1Er-0.1Zr alloys in different rolling process.

In this paper, the factors affecting the planar anisotropy of the alloy plane are discussed by two kinds of factors. First is the second-phase particles; the alloy is distributed with $\delta'$ ($Al_2Cu$) as the main reinforcing phase, which is a spherical substable phase with a cubic superlattice crystal structure ($L1_2$). Since the $\delta'$ phase is a spherical phase, its enhancement effect on the strength of the alloy is basically the same in different loading directions. However, the low shear modulus of the $\delta'$ phase enables the dislocations to repeatedly slide on the same slip surface and shear the $\delta'$ phase, which can easily lead to the anisotropy of the alloy elongation. The $\delta'$ phase acts with the dislocations by a cut-through mechanism during plastic deformation of the alloy, which promotes the coplanar slip, softens the slip surface gradually, and reduces the number of effective slip coefficients, which can easily lead to the localized stress concentration at the grain boundaries and cause brittle fracture in the alloy. Brittle fracture occurs [14,15]. This is the main reason that affects the plasticity change in the alloy in different directions. Second is with regard to the type of fabrication: the main changes in the type of fabrication of the alloy after cross-rolling are the elimination of brasslike fabrication {011}<111> and the enhancement of brass R-type fabrication {111}<112>. It is shown that the stronger the brasslike weave is, the higher the anisotropy of the alloy is and the higher the IPA index is, while the {111} surface weave can obviously inhibit the anisotropy of the alloy [13,32], which causes the alloy to be strongly anisotropic in the unidirectional rolling process, and a lower anisotropy can be obtained in the cross-rolling process.

## 4. Conclusions

(1) As the rolling deformation increases, the Cu-rich phase in the Al-3Li-1Cu-0.4Mg-0.1Er-0.1Zr alloy continues to fragment and dissolve, resulting in a continuous decrease in size from 30 μm to about 8 μm with a more uniform distribution.

(2) In the alloy with the rolling reduction of 90%, abundant $Al_3Li$ phases are precipitated, and an $Al_3$(Er, Zr, Li) core–shell composite phase is formed. Additionally, recrystallization occurs by the PSN mechanism.

(3) The strength and plasticity of Al-3Li-1Cu-0.4Mg-0.1Er-0.1Zr alloys increases with the increase in rolling reduction, in which the tensile strength of the alloy specimen with 90% of rolling deformation is 362MPa and the elongation is 19.1%.

(4) Compared with the normal rolled specimen, the cross-rolled specimen has relatively lower strength and elongation. The cross-rolling eliminates the brasslike texture {011}<111> generated by normal rolling and enhances the brass R-type texture {111}<112>.

(5) The IPA of the strength decreases from 10.1% in normal rolling to 5.5% in cross-rolling, and the particle stimulated nucleation of the elongation decreases from 12.8% to 3.3%. The cross-rolling obviously decreases the anisotropy.

**Author Contributions:** Z.M.: investigation, materials processing, and writing original draft. T.Z.: casting, heat treatment. D.S.: rolling processing. B.Q.: microstructure characterizations. N.T.: chemical composition designment. S.B.: texture characterization. R.W.: conceptualization, methodology, and revising original draft. All authors have read and agreed to the published version of the manuscript.

**Funding:** This research was funded by National Key Research and Development Program of China (grant number 2021YFE0103200), National Natural Science Foundation of China (grant numbers 52261135538, 51971071, U21A2049, 52271098), Fundamental Research Funds for the Central Universities (grant number 3072022QBZ1002), and the Russian Science Foundation (grant number 23-49-00098).

**Data Availability Statement:** The data that support the findings of this study are available from the corresponding author upon reasonable request.

**Conflicts of Interest:** The authors declare no conflict of interest.

## References

1. Jiang, L.; Rouxel, B.; Langan, T.; Dorin, T. Coupled segregation mechanisms of Sc, Zr and Mn at θ′ interfaces enhances the strength and thermal stability of Al-Cu alloys. *Acta Mater.* **2021**, *206*, 116634. [CrossRef]
2. Han, J.; Wang, H.; Xu, A.; Niu, K. Enhanced matrix precipitation of T1 ($Al_2CuLi$) phase in AA2055 Al–Li alloy during stress aging process. *Mater. Sci. Eng. A* **2021**, *827*, 142057. [CrossRef]
3. Wu, L.; Li, X.; Wang, H. Significant roles of minor Mg addition in microstructural evolution and mechanical properties of cast Al-3Li-2Cu-0.15 Zr alloy. *J. Alloys Compd.* **2021**, *857*, 157600. [CrossRef]
4. Abd El-Aty, A.; Xu, Y.; Guo, X.; Zhang, S.-H.; Ma, Y.; Chen, D. Strengthening mechanisms, deformation behavior, and anisotropic mechanical properties of Al-Li alloys: A review. *J. Adv. Res.* **2018**, *10*, 49–67. [CrossRef] [PubMed]
5. Booth-Morrison, C.; Dunand, D.C.; Seidman, D.N. Coarsening resistance at 400 °C of precipitation-strengthened Al–Zr–Sc–Er alloys. *Acta Mater.* **2011**, *59*, 7029–7042. [CrossRef]
6. Sun, Z.; Wang, H.; Tian, X.; He, B. Developing a novel lightweight Al–Mg–Li alloy for laser powder bed fusion additive manufacturing: Parameter optimization, microstructure evolution, and mechanical performance. *Mater. Sci. Eng. A* **2023**, *872*, 144992. [CrossRef]
7. Yang, Z.; Liao, Y.; Wu, R.; Sun, D.; Liu, M.; Wang, Y. Effect of Single-Pass Large-Strain Rolling on Microstructure and Mechanical Properties of Al-3Li-1Cu-0.2 Er-0.1 Zr Alloy. *J. Mater. Eng. Perform.* **2022**, *31*, 3287–3298. [CrossRef]
8. Zhang, P.; Chen, M. Progress in characterization methods for thermoplastic deforming constitutive models of Al–Li alloys: A review. *J. Mater. Sci.* **2020**, *55*, 9828–9847. [CrossRef]
9. Chung, T.F.; Lin, S.Q.; Li, H.; Hsiao, C.N.; Yang, J.R.; Tsao, C.S.; Ohmura, T. Microstructure evolution and shearing behaviour of $\delta''/\theta''/\delta''$ precipitates in an aged Al-Cu-Li-Mg aluminium alloy. *J. Alloys Compd.* **2023**, *953*, 170095. [CrossRef]
10. Li, Y.; Xu, G.; Peng, X.; Guo, G.; Liu, S.; Liang, X. Research on microstructure and mechanical properties of 2050 Al-Li alloy during non-isothermal aging. *J. Alloys Compd.* **2022**, *906*, 163977. [CrossRef]
11. Ye, F.; Mao, L.; Rong, J.; Zhang, B.; Wei, L.; Wen, S.; Jiao, H.; Wu, S. Influence of different rolling processes on microstructure and strength of the Al–Cu–Li alloy AA2195. *Prog. Nat. Sci. Mater. Int.* **2022**, *32*, 87–95. [CrossRef]
12. Zhuo, X.; Zhang, Q.; Liu, H.; Hu, Z.; Zhang, P.; Jiang, J.; Ma, A.; Wu, Y. Enhanced tensile strength and ductility of an Al-6Si-3Cu alloy processed by room temperature rolling. *J. Alloys Compd.* **2022**, *899*, 163321. [CrossRef]
13. Wang, Y.; Zhao, G.; Xu, X.; Chen, X. Effect of extrusion parameters on microstructure, texture and mechanical property anisotropy of spray deposited 2195 Al–Li alloy profile. *Mater. Sci. Eng. A* **2021**, *818*, 141406. [CrossRef]
14. Xu, X.; Hao, M.; Chen, J.; He, W.; Li, G.; Jiao, C.; Burnett, T.L.; Zhou, X. Influence of microstructural and crystallographic inhomogeneity on tensile anisotropy in thick-section Al–Li–Cu–Mg plates. *Mater. Sci. Eng. A* **2022**, *829*, 142135. [CrossRef]
15. Li, H.R.; Ma, Y.L.; Li, J.F.; Liu, D.Y.; Lu, D.D.; Guo, Y.J. Investigation of anisotropy and structure variation of spray-formed 2195 Al-Li alloy via final temperature-controlled rolling and cold rolling. *J. Alloys Compd.* **2023**, *937*, 168414. [CrossRef]
16. Huh, M.; Cho, S.; Engler, O. Randomization of the annealing texture in aluminum 5182 sheet by cross-rolling. *Mater. Sci. Eng. A* **2001**, *315*, 35–46. [CrossRef]
17. Nayan, N.; Mishra, S.; Prakash, A.; Murty, S.; Prasad, M.; Samajdar, I. Effect of cross-rolling on microstructure and texture evolution and tensile behavior of aluminium-copper-lithium (AA2195) alloy. *Mater. Sci. Eng. A* **2019**, *740*, 252–261. [CrossRef]
18. Li, Z.; Chen, L.; Tang, J.; Sun, W.; Zhao, G.; Zhang, C. Improving mechanical anisotropy and corrosion resistance of extruded AA7075 alloy by warm cross rolling and annealing. *J. Alloys Compd.* **2021**, *863*, 158725. [CrossRef]
19. Medjahed, A.; Moula, H.; Zegaoui, A.; Derradji, M.; Henniche, A.; Wu, R.; Hou, L.; Zhang, J.; Zhang, M. Influence of the rolling direction on the microstructure, mechanical, anisotropy and gamma rays shielding properties of an Al-Cu-Li-Mg-X alloy. *Mater. Sci. Eng. A* **2018**, *732*, 129–137. [CrossRef]

20. Abd El-Aty, A.; Xu, Y.; Zhang, S.-H.; Ha, S.; Ma, Y.; Chen, D. Impact of high strain rate deformation on the mechanical behavior, fracture mechanisms and anisotropic response of 2060 Al-Cu-Li alloy. *J. Adv. Res.* **2019**, *18*, 19–37. [CrossRef]
21. Zhong, F.; Wu, H.; Jiao, Y.; Wu, R.; Zhang, J.; Hou, L.; Zhang, M. Effect of Y and Ce on the microstructure, mechanical properties and anisotropy of as-rolled Mg-8Li-1Al alloy. *J. Mater. Sci. Technol.* **2020**, *39*, 124–134. [CrossRef]
22. Yu, T.; Li, B.; Medjahed, A.; Hou, L.; Wu, R.; Zhang, J.; Sun, J.; Zhang, M. Impeding effect of the $Al_3$ (Er, Zr, Li) particles on planar slip and intergranular fracture mechanism of Al-3Li-1Cu-0.1 Zr-X alloys. *Mater. Charact.* **2019**, *147*, 146–154. [CrossRef]
23. Shi, C.; Zhang, L.; Wu, G.; Zhang, X.; Chen, A.; Tao, J. Effects of Sc addition on the microstructure and mechanical properties of cast Al-3Li-1.5 Cu-0.15 Zr alloy. *Mater. Sci. Eng. A* **2017**, *680*, 232–238. [CrossRef]
24. Chen, A.; Wu, G.; Zhang, L.; Zhang, X.; Shi, C.; Li, Y. Microstructural characteristics and mechanical properties of cast Al-3Li-xCu-0.2 Zr alloy. *Mater. Sci. Eng. A* **2016**, *677*, 29–40. [CrossRef]
25. Mikhaylovskaya, A.; Mochugovskiy, A.; Levchenko, V.; Tabachkova, N.Y.; Mufalo, W.; Portnoy, V. Precipitation behavior of L12 Al3Zr phase in Al-Mg-Zr alloy. *Mater. Charact.* **2018**, *139*, 30–37. [CrossRef]
26. Xue, C.; Wang, S.; Zhang, Y.; Tian, G.; Yang, X.; Chang, X.; Ke, Y.; Xie, Z.; Wang, J. Uncovering the kinetics of Li-rich clusters and monodisperse core–shell $Al_3$ (Zr, Sc) structures in Al–Li–Cu alloys. *Mater. Sci. Eng. A* **2023**, *881*, 145393. [CrossRef]
27. Wang, Y.; Zhang, S.; Wu, R.; Turakhodjaev, N.; Hou, L.; Zhang, J.; Betsofen, S. Coarsening kinetics and strengthening mechanisms of core-shell nanoscale precipitates in Al-Li-Yb-Er-Sc-Zr alloy. *J. Mater. Sci. Technol.* **2021**, *61*, 197–203. [CrossRef]
28. Shen, K.; Timko, M.; Li, Y.-J.; Toal, R.; Santos, N.; Es-Said, S.; Ba Thaung, S.; Guevara, L.; Riebe, R.; Es-Said, O.S. The Effect of Temper, Grain Orientation, and Composition on the Fatigue Properties of Forged Aluminum-Lithium 2195 Alloy. *J. Mater. Eng. Perform.* **2019**, *28*, 5625–5638. [CrossRef]
29. Chen, X.; Zhao, G.; Liu, G.; Sun, L.; Chen, L.; Zhang, C. Microstructure evolution and mechanical properties of 2196 Al-Li alloy in hot extrusion process. *J. Mater. Process. Technol.* **2020**, *275*, 116348. [CrossRef]
30. Wang, X.-Y.; Jiang, J.-T.; Li, G.-A.; Wang, X.-M.; Shao, W.-Z.; Zhen, L. Particle-stimulated nucleation and recrystallization texture initiated by coarsened Al2CuLi phase in Al–Cu–Li alloy. *J. Mater. Res. Technol.* **2021**, *10*, 643–650. [CrossRef]
31. Fu, R.; Huang, Y.; Liu, Y.; Li, H. Microstructural Evolution and Dynamic Recrystallization Behavior of the Homogenized 2195 Al–Li Alloy During Hot Deformation. *Met. Mater. Int.* **2023**, *29*, 2605–2622. [CrossRef]
32. Wang, D.; Gao, C.; Luo, H.-Y.; Yang, Y.-H.; Ma, Y. Texture evolution behavior and anisotropy of 2A97 Al–Li alloy during recrystallization at elevated temperature. *Rare Met.* **2018**, *42*, 3139–3149. [CrossRef]
33. Guo, Y.-J.; Li, J.-F.; Lu, D.-D.; Dong, S.-Y.; Zeng, G.-J.; Ma, Y.-L.; You, W.; Chen, Y.-L.; Zhang, X.-H.; Zhang, R.-F. Characterization of Al3Zr precipitation via double-step homogenization and recrystallization behavior after subsequent deformation in 2195 Al-Li alloy. *Mater. Charact.* **2021**, *182*, 111549. [CrossRef]
34. Tangen, S.; Sjølstad, K.; Furu, T.; Nes, E. Effect of concurrent precipitation on recrystallization and evolution of the P-texture component in a commercial Al-Mn alloy. *Metall. Mater. Trans. A* **2010**, *41*, 2970–2983. [CrossRef]
35. Bennett, T.; Petrov, R.; Kestens, L.; Zhuang, L.-Z.; De Smet, P. The effect of particle-stimulated nucleation on texture banding in an aluminium alloy. *Scr. Mater.* **2010**, *63*, 461–464. [CrossRef]
36. Tian-Zhang, Z.; Long, J.; Yong, X.; Shi-Hong, Z. Anisotropic yielding stress of 2198 Al–Li alloy sheet and mechanisms. *Mater. Sci. Eng. A* **2020**, *771*, 138572. [CrossRef]

metals

*Article*

# The Stray Grains from Fragments in the Rejoined Platforms of Ni-Based Single-Crystal Superalloy

**Miao Huo [1,*], Chuyue Chen [1], Hangyue Jian [1], Wenchao Yang [2] and Lin Liu [2]**

[1] School of Materials Science and Engineering, Xi'an Shiyou University, Xi'an 710065, China
[2] State Key Laboratory of Solidification Processing, Northwestern Polytechnical University, Xi'an 710072, China
* Correspondence: huomiao8888@163.com

**Abstract:** Nickel-based single crystal superalloy is the most important material for blade preparation. However, some solidification defects inevitably occur during the process of preparing single-crystal blades through directional solidification. In this study, in order to study the origin of misorientation defects during solidification, a model with rejoined platforms was designed according to the geometry of single-crystal guide vanes. Electron Back-Scattering Diffraction (EBSD) was used to quantify the orientation deviation of the dendrites and identify the solidification defects in the rejoined platforms. The results showed that stray grain defects appeared in the platforms and their misorientation changed gradually, not abruptly. Combined with the simulation results, it was proposed that the stray grains formed as the result of the dendrites fragment, which was induced by solute enrichment in the mushy zone during solidification. Meanwhile, it was accompanied by a obvious dendritic deformation, which was caused by solidification shrinkage stress. This suggested that the fragmentation was induced by multiple factors, among which, the concave interface shape provided favorable conditions for solute enrichment, and the dynamic variability in the local thermal gradient and fluctuations of the solidification rate might play catalytic roles.

**Keywords:** Ni-based single superalloy; stray grains; fragments; simulation

## 1. Introduction

Due to the complex shapes of the turbine blades and guide vanes in advanced gas turbine systems, investment single-crystal (SX) casting is the preferred manufacturing route [1]. There are two prevalent techniques for fabricating SX blades: the grain selection technique and the seeding method. However, no matter which method is adopted, misorientation defects [2–4] are almost inevitable. Stray grain (SG) defects are common, particularly around the platforms and shroud regions [5–9] of the turbine blade, and can lead to component invalidation.

Therefore, some studies about the formation of stray grains in single-crystal superalloy have been carried out. In 1996, Meyer et al. [8] first depicted that SGs nucleated in a thermally undercooled zone caused by a macroscopically curved liquidus isotherm in the platform ends. In 1997, Bussac et al. [10] made a mathematical deduction and derived a criterion to predict SG formation at abrupt cross-sections, which was based on several assumptions rather than an actual casting process. Later, a lot of efforts were paid to investigating the influencing factors of SG formation: withdrawal rates and isothermal inclination angles [7], the undercoolabilities of superalloys [11], alloy composition [12], platform dimension [13], and grain orientation [14]. However, all these studies were based on the presumable fact that SGs are caused by heterogeneous nucleation. In fact, some SGs in platforms originate from dendrite deformation or dendrite fragmentation. In 2011, Zhou et al. [15] proposed that the occurrence of SGs at diverging boundaries was not caused by nucleation, but by the bending or detachment of side arms. In fact, as early as 1996, Meyer et al. [8] found a fine dentritic structure in the corners of a shroud, but they did not provide an explanation for this phenomenon. In 2009, Ma et al. [16] also found a similar dendritic structure of fine equiaxed grains in the overhanging parts of the shrouds with no heat conductor. They called them interdendritic micro equiaxed grains. Because the dendrite trunks were still "single crystal", they held a view that these grains could only damage the local monocrystality. Finally, they analyzed the formation mechanism of this defect and concluded that the dendrite roots in the undercooled melt were weak and could partially remelt during the subsequent coarsening process caused by the recalescence upon the sudden release of latent heat. Furthermore, Hao et al. [17] found a large number of misoriented grains between the primarily solidified dendrite stems in the specially shaped shroud during the directional solidification process. They attributed this phenomenon to the process of recalescence, which was caused by the accumulation of the solidification of latent heat. Wang et al. [18] also observed fine equiaxed grains in the interdendritic regions at the lowest shroud of single-crystal blades. They analyzed two formation mechanisms for this solidification defect. The first mechanism they analyzed was consistent with the viewpoint of Ma et al. [16]. They inferred that the undercooling of the melt provided a possibility for these grains, and positive segregation elements accumulating easily during transverse growth could have further refined the roots of the dendrite arms, some of which were remelted as a result of the sudden release of the latent heat of the recoalescence. The second formation mechanism they proposed was that the fine equiaxed grains appeared to be a fragmentation of the secondary dendrite arms caused by the contraction stress in the mushy zone during solidification. After that, Sun et al. [19] also discovered severe equiaxed grain defects in the platforms of a simplified turbine blade when studying dendrite deformation in Ni-based superalloys. However, they did not provide any explanation, possibly because there were too many factors affecting this type of defect. In summary, in Ni-based single-crystal superalloys, the formation of such defects might be related to latent heat, melt undercooling, or stress, but no consensus has been reached. In other alloy systems, in situ observation technology has been used to investigate the dendrite growth during the solidification process. The experimental results of a Sn-Bi alloy [20] and Al-20 wt.% Cu [21] revealed that the fragments could also be the source of an SG defect. The fragments were always accompanied by instability and solute segregation during the solidification process. It can be seen that such defects are common, but their formation mechanism is still unclear; thus, it is necessary to conduct further investigations.

At the present stage, it is quite difficult to conduct a real-time observation of the solidification process, dendrite growth, and dendrite fragment of superalloys, because of their high melting point. Therefore, very little is known about the factors that promote the SG defect from fragments during directional solidification. Therefore, the present study was carried out to explore the origin of these fragments through an experimental investigation and finite element simulation using ProCAST.

## 2. Experimental and Simulation Methods

### 2.1. Experiment Methods

A model consisting of 3 rejoined platforms (from bottom to top, labelled 1, 2, and 3, respectively) and transverse cylindrical bars with the same diameter (ϕ8 mm) and an equal length (52 mm) (Figure 1a) was used to fabricate single-crystal castings. The single-crystal castings were prepared using a modified Bridgman directional casting furnace. The furnace mainly consisted of a vacuum induction melting unit, a graphite heater unit, and a cooling zone (Figure 1b). During the process of directional solidification, the furnace chamber was evacuated and the ceramic mold was preheated. Molten DD6 (composition in Table 1) was poured into the mold and maintained for 10–20 min. Finally, the ceramic mold was withdrawn from the furnace at a controlled withdrawal rate of 150 μm/s.

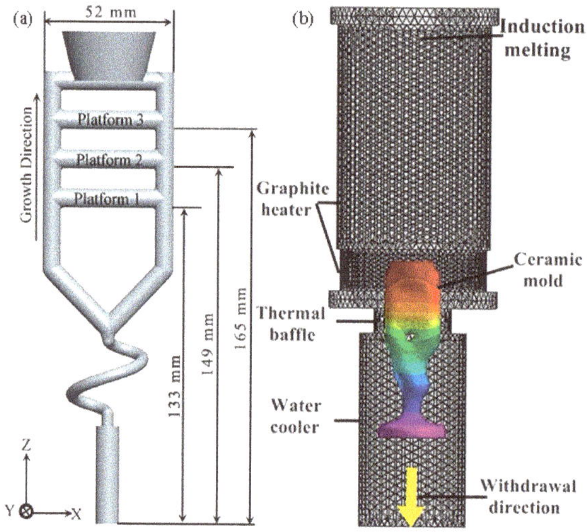


**Figure 1.** Schematic of the sample assembled with a spiral grain selector (**a**) and the FEM mesh model of the casting system (**b**).

**Table 1.** Nominal compositions of DD6.

| Ni | Cr | Co | Mo | W | Ta | Re | Nb | Al | Hf | C |
|---|---|---|---|---|---|---|---|---|---|---|
| Bal. | 4.3 | 9 | 2 | 8 | 7.5 | 2 | 0.5 | 5.6 | 0.1 | 0.006 |

After the casting process, the remnant mold was removed with sandpaper and the casting was cleared. Then, the platforms were sectioned from the sample longitudinally and transversely and prepared for a metallographic analysis. In order to reveal the microstructure, well-polished specimens were etched with a solution of $HNO_3$, HF, and $C_3H_8O_3$ (volume ratio 1:2:4). The microstructure was observed using optical microscopy (OM; Leica DM-4000 M, Leica, Berlin, Germany).

An Electron Back-Scattering Diffraction (EBSD) measurement was employed to determine the crystallographic orientations of the dendrites. The platform samples were mechanically polished to remove stress and obtain a flat surface. The EBSD measurements were performed in ZEISS SUPRA 55 SEM (Carl Zeiss, Inc., Oberkochen, Germany) equipped with an HKL system and Channel 5 analysis software (Oxford instruments, Oxford, UK). During the EBSD scanning process, a step size of 40 μm was used for the EBSD mapping. All-Euler maps and pole figures of the EBSD results were obtained using the Channel 5 analysis software.

### 2.2. Simulation Methods

The castings were simulated using a finite element analysis with ProCAST in order to exhibit the thermal profile that evolved during the directional solidification process. The solidus and liquidus temperatures of the alloy were determined to be 1342 and 1399 °C, respectively, using a differential scanning calorimetry (DSC) analysis. The simulation was initiated before the withdrawal; the melt and chill-plate temperatures were set to 1550 and 400 °C, respectively. In order to make the simulation results as accurate as possible, the meshing size was set to 1 mm. The boundary conditions, initial conditions, and interfacial heat transfer coefficients were listed in reference [22].

Such simulations are relatively routine. The salient points of the calculations are highlighted below:

(1) The simulation models of the castings were constructed according to the actual dimensions. The thickness of the ceramic shell was assigned to be 7 mm from the actual measurement.
(2) The heat transfer was assumed to be heat conduction and radiation, and heat convection was ignored. The coefficients of the thermal emissivity and thermal conductivity were gained from reference [23–26] or from the experiments. The thermophysical parameters were calculated using the commercial software JMatPro (Trade Mark of Sente Software Ltd. Surrey Technology Center, Surrey GU2 7YG, UK). The thermal results were validated through a thermocouple instrumented experiment.

## 3. Results

### *3.1. Dendrite Evolution in the Rejoined Platform*

The dendritic structure of platform 3 in the XY plane is shown in Figure 2. Compared to platforms 1 and 2, the characteristics of the defects in platform 3 were most obvious, which was taken as an example to investigate the formation of stray grains in the rejoined platforms. For ease of expression, the platform was divided into three regions, marked as left side (L), middle region (M), and right side (R). According to the growth morphology of the dendrites, the platform was divided into three regions: I, II, and III.

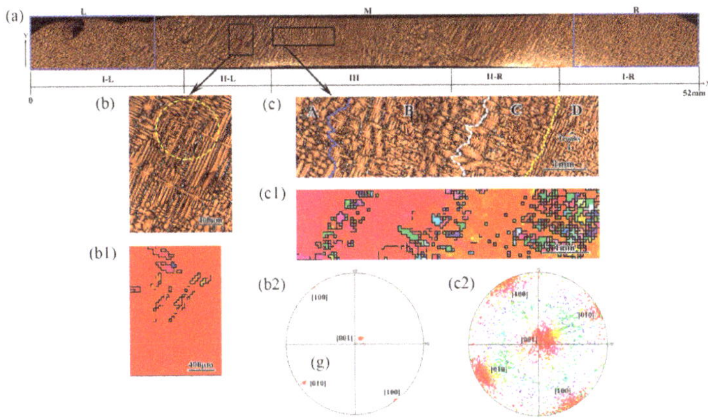


**Figure 2.** Transversal (parallel to plane XY) microstructures of the platform 3 (**a**), enlarged view of the stray grains in the region M (**b**,**c**), their EBSD scans (**b1**,**c1**), and their polar diagrams (**b2**,**c2**).

In the regions L and R, the original primary dendrite arms grew regularly and the secondary dendrites formed a two-dimensional cruciform structure. With the dendrites growing into region M, the cruciform dendrites remained until the end of region I (I-L and I-R). Starting from region II (II-L and II-R), long secondary dendrites formed on the outside of the platform. This was because of the lateral growth of the secondary dendrites caused by lateral heat dissipation. Furthermore, some misorientation grains occurred in region II, and their size was comparable with that of the surrounding tertiary dendrites, as shown in the enlarged view in Figure 2b and its EBSD scan in Figure 2(b1). As seen the pole figure in Figure 2(b2), the orientations of these grains were close, all deviating, but being distributed around [001]. When the dendrites continued to grow into region III, the long secondary branches still existed but became thinner, and a lot of grains with various sizes formed, as shown in the enlarged view in Figure 2c.

Combining the metallograph in Figure 2c and its EBSD scan in Figure 2(c1), it can be seen that there were four stages in region III: (1) in region A, there were no broken dendrites, but thinner secondary branches; (2) in region B, small-scale broken tertiary dendrite arms appeared first, followed by a stable region, and the orientation of the trunks was unchanged, as shown in the red region in Figure 2(c1); (3) in region C, the broken branches appeared for the second time, and were also followed by a stable region. However, there was a slight deviation in their orientation, as shown in the orange region in Figure 2(c1); and (4) in region D, almost all the tertiary dendrite arms detached from their parent dendrites, and their orientation was deviated according to the pole figure in Figure 2(c2), in which the corresponding red color gradually turned orange. This meant that there were many broken branches and their orientation had seriously deviated from the main trunks. From then on, the integrity of the single crystal was destroyed completely. There was no doubt that these misorientation grains were stray grains.

From the EBSD scan in Figure 2c, it is shown that the orientations of these detached tertiary grains were different from each other, but they were from the same trunks (the red or orange area in Figure 2c). Meanwhile, the pole figure in Figure 2(c2) displayed that the orientation distribution of the test region was dispersed, but around the trunks (close to the [001] direction), and only a few grains deviated seriously, as shown by the green and blue dots in Figure 2(c2). In addition, the orientation transition of the trunks was from the red to orange position, which indicated that the long secondary dendrites could have been deformed. In other words, the orientation deviation was gradual rather than abrupt. Thus, it can be seen that the secondary dendrites deformed during growth, then their derived tertiary dendrites detached from the mother branches and underwent local torsion, resulting in disorientation.

In order to further confirm the dendrites' evolution, the formation of fragments, and the dendrite deformation in the rejoined platforms, a metallographic test and EBSD measurements were performed on the longitudinal section (parallel to plane XZ) of platform 3, as shown in Figure 3. Similar to the dendritic evolution within the transversal surface, in regions L and R, the primary dendrites were arranged neatly like branches. As dendrites gradually grew into the middle region of the platform, the arrangement of the dendrites deteriorated. At first, there was a small number of detached dendrites, but later, a large number of broken dendrites appeared. In other words, the number of broken branches significantly increased as the dendrites gradually grew into the platform. Beyond that, another phenomenon, dendrite deformation, was found in the longitudinal section of the platform, as shown in Figure 3a. The red dashed line in Figure 3a is the growth path of the deformed dendrites. Figure 3(a1) and Figure 3(a2) are the EBSD scan image and polar diagram at the corresponding position, respectively. From these figures, it can be seen that the orientation of the main trunks gradually changed, corresponding to the color region in the figures from purple to yellow. This further verifies the existence of dendritic deformation. Finally, a grain boundary formed at the convergence interface, as shown by the yellow dashed line in Figure 3b and its EBSD scan (Figure 3(b1)). However, the grain boundary was closer to the left side of the platform. This might have been related to the original orientation of the dendrites on both sides and the competitive growth between them. In addition, all the fragments distributed along the deformed dendrite (Figure 3a) were located at the intersection, as shown in Figure 3c and its EBSD scan in Figure 3(c1).

### *3.2. Finite Element Simulation*

Based on the numerical simulation, the shapes of the isotherm in the mushy zone of the rejoined platform with the advancing of the solidification interface were predicted, as shown in Figure 4. It was found that the shapes of the isotherm varied at different times. Firstly, the temperature on both sides of the platform was lower than liquidus (1342 °C) and the interface slightly tilted towards the platform, as shown in Figure 4a. Further withdrawal of the mold resulted in shifting the liquidus isotherm into the platform, as shown in Figure 4b. Then, the interface became a symmetrical sloping groove, as shown in Figure 4c,d. This meant that the lateral heat extraction was enhanced in the platform and this inference was confirmed by the simulation results of the thermal gradient (Figure 5). Finally, directional solidification was completed on the top side of the platform, as shown in Figure 4e,f.

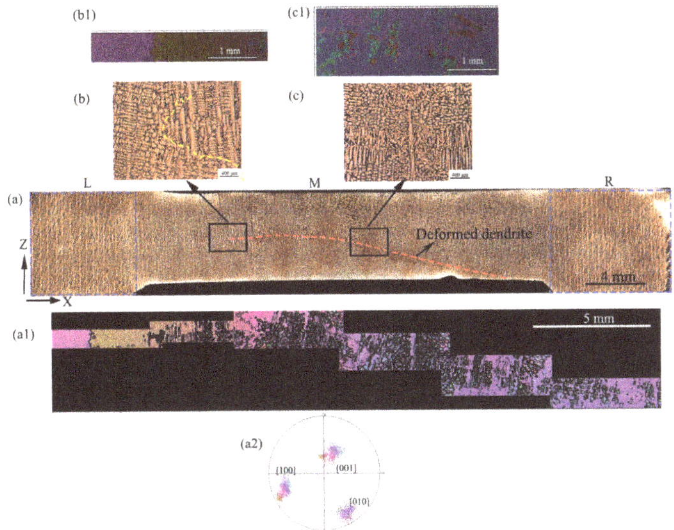


**Figure 3.** Longitudinal (parallel to plane XZ) microstructure of the platform 3 (**a**), EBSD scan of the region with fragments (**a1**), and its polar diagram (**a2**), enlarged views of different regions (**b**,**c**), and their corresponding EBSD scans (**b1**,**c1**).

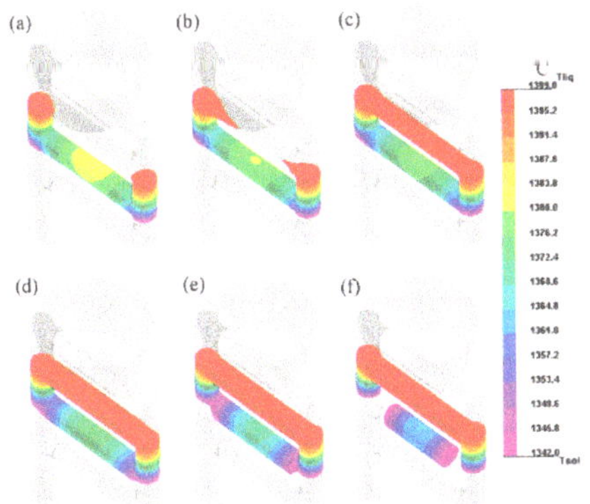


**Figure 4.** The shape of isotherm in mushy zone of the rejoined platform with the advancing of solidification interface, (**a**) t = 1014 s, (**b**) t = 1024 s, (**c**) t = 1034 s, (**d**) t = 1044 s, (**e**) t = 1054 s, (**f**) t = 1064 s.

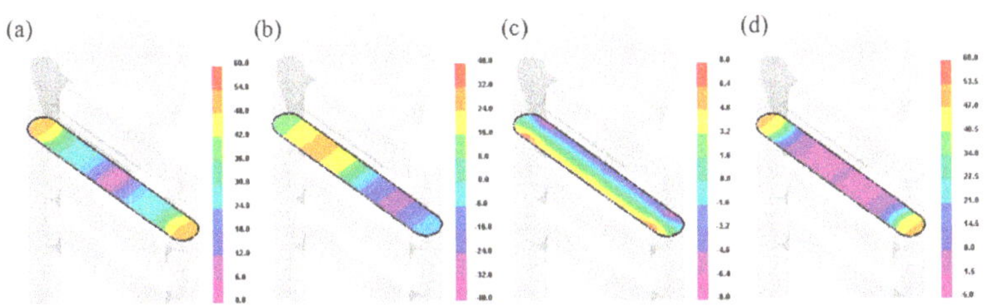


**Figure 5.** The total thermal gradient (**a**) and components in three directions of the rejoined platform (**b**) lateral thermal gradient (along X direction) and (**c**) along Y direction, (**d**) axial thermal gradient (along Z direction) (°C/cm) [27].

Figure 5 shows the simulation results of the total thermal gradient and the components in three directions of the rejoined platform. As seen from this figure, the total thermal gradient gradually decreased from both sides to the middle region (Figure 5a), the thermal gradients along directions X and Y fluctuated (Figure 5b,c), and the thermal gradient along the Z direction was the lowest in the middle region of the platform (Figure 5d). Thus, it can be seen that the lateral thermal gradients (directions X and Y) were higher but fluctuating. The simulation results of the solidification rate also exhibited similar laws, as shown in Figure 6.

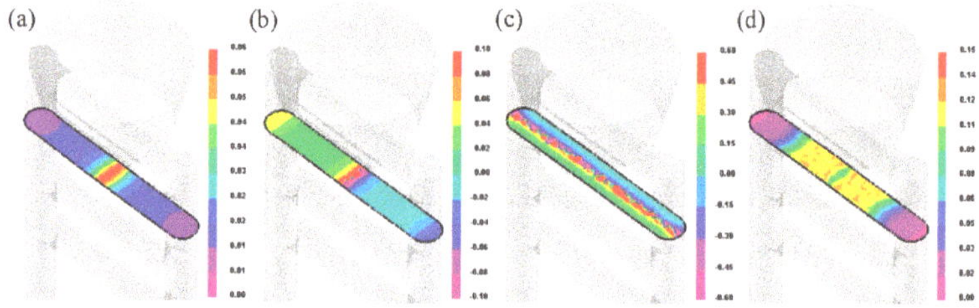


**Figure 6.** Contour plots of total solidification rate (**a**) and the solidification rates in X direction (**b**), in Y direction (**c**), and in Z direction (**d**).

## 4. Discussion

Based on the above metallographic images, the existence of dendrite trunks meant that initial nucleation could be excluded. Further analysis of the EBSD test results showed that the broken branches in the middle region of the rejoined platforms could have been the result of dendrite fragmentation. The formation mechanism of these defects is discussed below.

As we know, the formation of microstructure and all potential defects during directional solidification are dependent on the local thermal condition, specifically, the thermal gradient and solidification rate [9,28]. Furthermore, the solute diffusion at the solid–liquid interface is also a critical factor affecting the evolution of dendritic structures [29–31]. In other words, there are three most concerned factors in the analysis of dendrite fragmentation: the thermal gradient, solidification rate, and solute enrichment. Meanwhile, previous studies have demonstrated that this could not involve a significant detachment of dendrites under steady-state conditions of growth rate and temperature gradient [32], but detachment could be induced with fluctuations in the solidification rate, particularly with deceleration [20,33], possibly preceded by acceleration [34]. Actually, dendrite growth is a three-dimensional process, and a change in growth condition in three directions (X, Y, and Z) is not synchronous, caused by the complex geometry. Therefore, fluctuations of the thermal conditions in any direction can lead to differences in microstructure and dendritic morphology.

In this experiment, the developed secondary dendrites in the platform were caused by good lateral heat dissipation conditions, which could be seen from the simulation results for the thermal gradient in Figure 5. After that, the local detached dendrites occurred firstly in region II (Figure 2b), and finally, a large amount of broken dendrites appeared in region III (Figure 2c), which could be divided into four stages. Based on the EBSD results, these defects could be classified as stray grains. In fact, their formation mechanisms were very complex and different in each region. Therefore, these should be discussed, respectively.

In region II, the total thermal gradient (Figure 5a) and total solidification rate (Figure 6a) were approximately constant. This meant that the local thermal condition was steady and was detrimental to the detachment of the dendrites. However, there were local tertiary dendrite fragments on the long secondary branches, as shown in Figure 2b. As mentioned earlier, obvious lateral growth was induced by the severe lateral heat extraction, and dendrites can grow regularly under stable conditions. In this sample, the inclined solid–liquid interface (Figure 4) and high lateral temperature gradient (Figure 5) in the platform contributed to the inevitable lateral growth in region II (II-L and II-R in Figure 2a). As the solidification front advanced, the secondary arms with severe lateral growth grew, ripened, and coarsened. During the earlier stages of the ripening, the roots or nodes became necked because of excess radial solute rejection from all the surrounding branches. As demonstrated in previous studies [35], fragmentation can occur naturally during coarsening or due to recalescence from eutectic solidification. Consequently, the tertiary branches around the crisscross might have been subjected to an enrichment of the solute in the pockets and detached from their attaching roots first, as shown by the fragments in the yellow dashed circle in Figure 2b. In fact, this could also have been the main reason for the distribution characteristics of the fragments in Figure 3. Thus, the formation mechanism of the broken branches in this region was similar to the research results of Ma et al. [16] and Wang et al. [18]. On the one hand, the roots of the tertiary dendrites were weak during the coarsening process of the surrounding dendrites, and on the other hand, the recalescence upon the sudden release of latent heat might have played a very important role, which ultimately led to small-scale fragments.

In region III, the thermal gradient (Figure 5a,b) and solidification rate (Figure 6a,b) fluctuated many times. As is known, each fluctuation would cause changes in the dendrite spacing and there was a transition for the dendrites to adjust. Previous studies have shown that this adjustment in dendrite spacing may lead to a lag in the thermal response, which might cause local solute pileup and eventually lead to crystal fragments [21,35,36]. This is why there were three regions of secondary dendrite fragments in the platform, as shown regions B, C, and D in Figure 2c. Furthermore, the shape of these regions was close to the solidification interface. As the solidification interface continued to move forward, the secondary branches grew regular again, as shown by the trunks in Figure 2c. However, due to the increase in the solidification rate in the X direction (Figure 5b), the secondary branches here were thinner compared to region II. Meanwhile, an accumulation of the solute was reinforced continuously, which would cause the solidus temperature to decrease [37]. Furthermore, the local torsion of the detached branches could be related to the stress that originated from the solidification process [38]. Consequently, almost all the tertiary arms (Figures 2c and 3c) detached from their parent branches. However, fragmentation occurred in the mushy zone, when the dendritic trunks had already formed, and the broken branches had no space to move or drift away, so only deflected or rotated around themselves. Thus, their orientations were random but their positions were unchanged. Eventually, the mosaic structure was retained in the solidification microstructure. As for the dendrite deformation, this was caused by the solidification shrinkage stress, as analyzed by the author in a previous study in reference [27]. Recently, Yang et al. [39] combined the cellular automaton-finite volume approach with the displacement-based finite element method to simulate dendrite growth and flow-induced deformation in an Al-4.5 wt%Cu alloy, in order to reveal the stress evolution during dendritic growth. It was found that dendrites could undergo visible mechanical bending under stress, which further confirms our speculation and analysis on dendritic deformation.

## 5. Conclusions

In the present research, the origin of stray grains in rejoined platforms was investigated using experiments and the finite element method. In Ni-based single-crystal superalloy, the stray grains from dendritic fragmentation were first determined and their possible formation mechanism was discussed. The reasons for the formation of the fragments were not entirely the same in different regions, as there were differences in the thermal fields and growth conditions of the dendrites. In general, the detachment of the tertiary arms was mainly caused by the solute enrichment at their roots and the latent heat, while the secondary fragments might have been related to the lag of the dendrite spacing adjustment caused by the fluctuation of the temperature field (especially the temperature gradient and solidification rate), with solute enrichment being the most fundamental reason for this.

Therefore, the fragments had different characteristics in different regions with fluctuations. Additionally, dendritic deformation defects were quite common in the rejoined platform, which were caused by solidification shrinkage stress. These defects would damage the integrity and properties of single crystals. There may be a risk of growth during heat treatment, and this would further damage the properties of the alloy. Therefore, their growth mechanism and control measures will be the focus of our attention in the future. Based on previous research results, measures such as controlling the solidification parameters, adjusting the casting structure, and adding auxiliary structures may make progress in reducing solidification defects.

**Author Contributions:** Conceptualization, L.L., W.Y. and M.H.; methodology, W.Y. and M.H.; validation, C.C., H.J. and M.H.; investigation, C.C., H.J. and M.H.; resources, L.L., W.Y. and M.H.; data curation, W.Y. and M.H.; writing—original draft preparation, M.H.; writing—review and editing, M.H.; funding acquisition, M.H. All authors have read and agreed to the published version of the manuscript.

**Funding:** This research was funded by National Natural Science Foundation of China (521001130); Natural Science Basic Research Program of Shaanxi (2021JQ-604, 2021JM-403); Scientific Research Program of Shaanxi Provincial Department of Education (21JC027).

**Data Availability Statement:** The raw/processed data required to reproduce these findings cannot be shared at this time due to technical or time limitations. They will be shared upon request.

**Conflicts of Interest:** The authors declare no conflict of interest.

## References

1. Reed, R.C. *The Superalloys Fundamentals and Applications*; Cambridge University Press: Cambridge, UK, 2006.
2. Wang, Z.; Li, J.; Liu, S.; Zhao, J.; Wang, X.; Yang, W. Investigation on freckle formation of nickel-based single crystal superalloy specimens with suddenly reduced cross section. *J. Alloys Compd.* **2022**, *918*, 165631. [CrossRef]
3. Yang, Q.; Zhu, X.; Wang, F.; Ma, D.; Wu, J. A study of sliver in C-shaped grain selectors during investment casting of single-crystal superalloy. *Metals* **2023**, *13*, 1102. [CrossRef]
4. Xia, H.; Yang, Y.; Feng, Q.; Xu, Q.; Dong, H.; Liu, B. Generation mechanism and motion behavior of sliver defect in single crystal Ni-based superalloy. *J. Mater. Sci. Technol.* **2023**, *137*, 232–246. [CrossRef]
5. Yang, Z.; Liu, C.; Hu, S.; Zheng, S.; Luo, Y.; Dai, S. Influence of platform position on stray grain nucleation in Ni-based single-crystal dummy blade clusters. *China Foundry* **2021**, *18*, 442–449.
6. Liu, Z.; Miao, K.; Lian, W.; Lu, Z.; Li, D. Elimination of the stray grain defects of single crystal blade by variable wall thickness sased on integral ceramic mold. *Metals* **2022**, *12*, 1832.
7. Ren, N.; Li, J.; Wang, B.; Zeng, L.; Xia, M.; Li, J. Design of variable withdrawal rate for superalloy single-crystal blade fabrication. *Mater. Des.* **2021**, *198*, 109347. [CrossRef]
8. Vehn, M.M.T.; Dedecke, D.; Paul, U.; Sahm, P.R. Undercooling related casting defects in single crystal Ni-based superalloy. In *Superalloys 1996*; Kissinger, R.D., Deye, D.J., Anton, D.L., Cetel, A.D., Nathal, M.V., Pollock, T.M., Woodford, D.A., Eds.; TMS: Pittsburgh, PA, USA, 1996; pp. 471–479.
9. Wang, W.; Kermanpur, A.; Lee, P.D.; McLean, M. Simulation of dendritic growth in the platform region of single crystal superalloy turbine blades. *J. Mater. Sci.* **2003**, *38*, 4385–4391.
10. Bussac, A.D.; Gandin, C.A. Prediction of a process window for the investment casting of dendritic single crystals. *Mater. Sci. Eng. A* **1997**, *237*, 35–42. [CrossRef]
11. Ma, D.; Xu, W.; Xu, F.; Wei, J.; Zhang, H. Comparative investigation of the undercooling capacity and single-crystal castability of some Ni-based superalloys. *Crystals* **2023**, *13*, 57. [CrossRef]
12. Zhang, X.; Zhou, Y.; Jin, T.; Sun, X.F. Study on the tendency of stray grain formation of Ni-based single crystal superalloys. *Acta Metall. Sin.* **2012**, *48*, 1229–1236. [CrossRef]
13. Meng, X.B.; Li, J.G.; Chen, Z.Q.; Wang, Y.H.; Zhu, S.Z.; Bai, X.F.; Wang, F.; Zhang, J.; Jin, T.; Sun, X.F.; et al. Effect of platform dimension on the dendrite growth and stray grain formation in a Ni-base single-crystal superalloy. *Metall. Mater. Trans. A* **2013**, *44*, 1955–1965. [CrossRef]
14. Li, Y.; Liu, L.; Huang, T.; Wang, H.F.; Zhang, J.; Fu, H.Z. Multi-scale characterization of stray grain in the platform of nickel-base single crystal turbine blade. *Vacuum* **2016**, *131*, 181–187.
15. Zhou, Y. Formation of stray grains during directional solidification of a nickel-based superalloy. *Scr. Mater.* **2011**, *65*, 281–284.
16. Ma, D.; Bührig-Polaczek, A. Application of a heat conductor technique in the production of single-crystal turbine blades. *Metall. Mater. Trans. B* **2009**, *40*, 738–748. [CrossRef]
17. Hao, H.; Jiang, W.; Xie, G.; Zhang, G.; Lu, Y.; Zhang, J.; Lou, L. Microstructure and grain orientation evolution of a specially shaped shroud during directional solidification process. *Prog. Nat. Sci. Mater. Int.* **2013**, *23*, 211–215. [CrossRef]

18. Wang, F.; Wu, Z.; Huang, C.; Ma, D.; Jakumeit, J.; Bührig-Polaczek, A. Three-dimensional dendrite growth within the shrouds of single crystal blades of a Nickel-based superalloy. *Metall. Mater. Trans. A* **2017**, *48*, 5924–5939. [CrossRef]
19. Sun, D.; Liu, L.; Huang, T.; Yang, W.; Li, Y.; Yue, Q.; Zhang, J.; Fu, H. Insight of the dendrite deformation in Ni-based superalloys for increased misorientation along convergent boundaries. *Prog. Nat. Sci. Mater. Int.* **2018**, *28*, 489–495. [CrossRef]
20. Yasuda, H.; Ohnaka, I.; Kawasaki, K.; Sugiyama, A.; Ohmichi, T.; Iwane, J.; Umetani, K. Direct observation of stray crystal formation in unidirectional solidification of Sn Bi alloy by X-ray imaging. *J. Cryst. Growth* **2004**, *262*, 645–652.
21. Mathiesen, R.H.; Arnberg, L. Stray crystal formation in Al-20wt.% Cu studied by synchrotron X-ray video microscopy. *Mater. Sci. Eng. A* **2005**, *413–414*, 283–287.
22. Huo, M.; Liu, L.; Yang, W.; Sun, D.; Hu, S.; Zhang, J.; Fu, H. Formation of slivers in the extended cross-section platforms of Ni-based single crystal superalloy. *Adv. Eng. Mater.* **2018**, *20*, 1701189.
23. Miller, J.D.; Pollock, T.M. The effect of processing conditions on heat transfer during directional solidification via the bridgman and liquid metal cooling processes. *Metall. Mater. Trans. A* **2014**, *45*, 411–425. [CrossRef]
24. Gao, S.F.; Liu, L.; Wang, N.; Zhao, X.B.; Zhang, J.; Fu, H. Grain selection during casting Ni-base, single-crystal superalloys with spiral grain selector. *Metall. Mater. Trans. A* **2012**, *43*, 3767–3775.
25. Wang, N.; Liu, L.; Gao, S.; Zhao, X.; Huang, T.; Zhang, J.; Fu, H. Simulation of grain selection during single crystal casting of a Ni-base superalloy. *J. Alloys Compd.* **2014**, *586*, 220–229. [CrossRef]
26. Brundidge, C.L.; Miller, J.D.; Pollock, T.M. Development of dendritic structure in the liquid-metal-cooled, directional-solidification process. *Metall. Mater. Trans. A* **2011**, *42*, 2723–2732. [CrossRef]
27. Huo, M.; Liu, L.; Yang, W. Dendrite deformation in the rejoined platforms of Ni-based single crystal superalloys. *Adv. Eng. Mater.* 2023; *early view*. [CrossRef]
28. Miller, J.D.; Pollock, T.M. Stability of dendrite growth during directional solidification in the presence of a non-axial thermal field. *Acta Mater.* **2014**, *78*, 23–36. [CrossRef]
29. Zhang, Y.; Zhou, J.; Yin, Y.; Ji, X.; Shen, X.; Guo, Z. Study on the solutal convection during dendrite growth of superalloy under directional solidification condition. *J. Mater. Res. Technol.* **2023**, *23*, 3916–3927. [CrossRef]
30. Zhang, H.; Wu, M.; Rodrigues, C.M.G.; Ludwig, A.; Kharicha, A.; Rónaföldi, A.; Roósz, A.; Veres, Z.; Svéda, M. Dendrite fragmentation mechanism under forced convection condition by rotating magnetic field during unidirectional solidification of AlSi7 alloy. *Acta Mater.* **2022**, *241*, 118391.
31. Ren, N.; Panwisawas, C.; Li, J.; Xia, M.; Dong, H.; Li, J. Solute enrichment induced dendritic fragmentation in directional solidification of nickel-based superalloys. *Acta Mater.* **2021**, *215*, 117043. [CrossRef]
32. Hansen, G.; Liu, S.; Lu, S.; Hellawell, A. Dendritic array growth in the systems $NH_4Cl$-$H_2O$ and $[CH_2CN]_2$-$H_2O$: Steady state measurements and analysis. *J. Cryst. Growth* **2002**, *234*, 731–739. [CrossRef]
33. Liu, S.; Lu, S.Z.; Hellawell, A. Dendritic array growth in the systems $NH_4Cl$-$H_2O$ and $[CH_2, CN]_2$-$H_2O$: The detachment of dendrite side arms induced by deceleration. *J. Cryst. Growth* **2002**, *234*, 740–750. [CrossRef]
34. Peng, P.; Li, X.Z.; Li, J.G.; Su, Y.Q.; Guo, J.J.; Fu, H.Z. Detachment of secondary dendrite arm in a directionally solidified Sn-Ni peritectic alloy under deceleration growth condition. *Sci. Rep.* **2016**, *6*, 27682. [CrossRef] [PubMed]
35. Li, B.; Brody, H.D.; Kazimirov, A. Real-time observation of dendrite coarsening in Sn-13%Bi alloy by synchrotron microradiography. *Phys. Rev. E* **2004**, *70*, 062602. [CrossRef] [PubMed]
36. Kurz, W.; Rappaz, M.; Trivedi, R. Progress in modelling solidification microstructures in metals and alloys. Part II: Dendrites from 2001 to 2018. *Int. Mater. Rev.* **2021**, *66*, 30–76. [CrossRef]
37. Yang, W.; Yu, H.; Wang, J.; Cai, C.; Xu, Z.; Li, S.; Liu, F.; Yang, G. Application of dendrite fragmentation to fabricate the homogeneous dispersed structure in undercooled Cu-Co immiscible alloy. *J. Alloys Compd.* **2011**, *509*, 9675–9678. [CrossRef]
38. Liu, F.; Cai, Y.; Guo, X.F.; Yang, G.C. Structure evolution in undercooled DD3 single crystal superalloy. *Mater. Sci. Eng. A* **2000**, *291*, 9–16.
39. Yang, L.; Ren, N.; Panwisawas, C.; Li, J.; Xia, M.; Dong, H.; Li, J. Melt flow-induced mechanical deformation of dendrites in alloy solidification: A coupled thermal fluid—Solid mechanics approach. *J. Mater. Res. Technol.* **2023**, *25*, 4094–4109. [CrossRef]

metals

*Article*

# Cracking Behavior and High-Temperature Thermoplastic Analysis of 09CrCuSb Steel Billets

**Zhixian Peng [1,*], Tao Mei [1], Jian Zheng [2], Yuan Yuan [2] and Liwang Wang [3]**

1 State Key Laboratory of Refractory Materials and Metallurgy, Wuhan University of Science and Technology, Wuhan 430081, China
2 Hunan Hualing Xiangtan Iron and Steel Co., Ltd., Xiangtan 411101, China
3 Zhejiang Jincheng New Material Co., Ltd., Changxing 313100, China; wanglw5133@163.com
* Correspondence: pengzhixian@wust.edu.cn

**Abstract:** This paper characterizes the surface crack morphology and elements of 09CrCuSb steel continuous casting slabs and studies the high-temperature thermo-plasticity of continuous casting slabs using a Gleeble 3500 thermal simulator. Combining the macro/micro fracture morphology and microstructure characteristics, the formation rules of slab cracks are discussed. The research results show that the increase in Ae3 temperature caused by changes in alloy elements results in the precipitation of a thin, film-like coexisting ferrite along the grain boundaries at a certain temperature, as well as the element segregation behavior of low-melting-point alloy elements at the original austenite grain boundaries, which are the main factors inducing cracks in 09CrCuSb steel continuous casting billets. The plasticity of 09CrCuSb steel at high temperatures is poor at 800~900 °C. In continuous casting process control, it is necessary to try to avoid long-term stay of the billet at this temperature range.

**Keywords:** 09CrCuSb; high-temperature thermoplastic; continuous casting billet cracking; boiler steel

## 1. Introduction

09CrCuSb steel, also known as ND steel, is widely used in equipment and devices that require resistance to low-temperature dew point corrosion of sulfur-containing flue gas due to its excellent performance in this area [1,2]. According to the new national energy industry standard "Technical Requirements for Steel Tubes for Boilers and Heat Exchangers NB/T47019-2021" implemented on 1 July 2021 in China [3], significant changes have been made to the composition of 09CrCuSb steel, particularly with a substantial increase in elements such as Cu, Sb, Sn, Mo, and W. The significance of the above elements' addition lies in the improvement of dew point corrosion resistance of sulfur-containing flue gas [4,5]. However, this has resulted in severe surface cracks in the continuous casting process, which not only increases the cost of repairs but also affects the quality of subsequent rolled products [6–9].

Many scholars have conducted research on the cracking of continuous casting billets of 09CrCuSb steel. For example, from the aspect of the cast process, Hu et al. [10] analyzed the crack morphology at the corners of 09CrCuSb steel and combined it with finite element simulation to analyze the temperature distribution in the second cooling zone during casting, determining that the cracks occurred mainly in the secondary water cooling zone during the casting process. Based on a 2D heat transfer model's real-time temperature field analysis for the entire slab section and process, Chen et al. [11] summarized that reducing or increasing the water flux in secondary cooling zones by over 20% and 80%, respectively, can maintain good thermoplastic properties in the straightening zone and can lower the risk of corner transverse crack formation. Furthermore, it should be noted that due to the high content of easy segregation elements at the grain boundaries, such as Cu, Sb, Sn, etc., in the alloy composition of 09CrCuSb steel, the problem of interface weakening caused by grain boundary segregation at high temperatures should also be taken seriously [12,13]. Chen et al. [6] studied the mechanism of surface crack formation in 09CrCuSb steel and found that surface crack defects were caused by slag oxidation and Cu element enrichment. In a similar study, Zhou et al. [14] found that Cu is hard to move from the grain boundary to the steel matrix in continuous casting due to its solubility being high and the diffusion coefficient being low in Fe. This leads to weakened grain boundaries that can cause cracks when the billet is bent or straightened [15,16]. Lu [17] analyzed the elements near the cracks of 09CrCuSb steel castings and found that there was a significant correlation between the presence of Cu and Sb precipitation phases and crack initiation. Huang [18] found that when the segregation of residual elements Sn and Sb on the grain boundaries was high, the thermal plasticity of the material was poor. Therefore, under the new national standard composition requirements, the research on the cracking of continuous casting billets of 09CrCuSb steel should further explore the distribution of crystal boundary elements and the process parameters, understand the micro-mechanism of billet cracking, and propose corresponding solutions.

This article focuses on the continuous casting billets of 09CrCuSb steel and analyzes and discusses the causes of cracking from the perspectives of high-temperature phase transformation and grain boundary segregation of elements. The recommendations for improving the process are also included based on thermal simulation, optical microscopy (OM), electron probe microanalysis (EPMA), and scanning electron microscopy + energy-dispersive X-ray spectroscopy (SEM + EDS).

## 2. Materials and Experimental Procedure

This experiment focuses on the preparation of 09CrCuSb steel continuous casting billets using the composition specified in the NB/T47019-2021 [3] national standard. The chemical composition is shown in Table 1. The continuous casting process has a casting cross-sectional size of 280 mm × 280 mm, a maximum overheat temperature of 30 °C in the tundish, a pulling speed of 0.95 m/min, and a measured temperature at the billet surface of 998 °C and at the corners of 916 °C during casting. After passing through the straightening machine, the measured temperatures at the billet surface and corners were 910 °C and 805 °C, respectively. After the casting process, the red billet showed no visible defects such as cracks or scratches, but macroscopic cracks were present on the surface of the continuous casting billet.

**Table 1.** Composition analysis of experimental steels (wt. %).

| | | | | | |
|---|---|---|---|---|---|
| 09CrCuSb | C% | P% | S% | Cu% | Cr% |
| | 0.06 | 0.009 | 0.003 | 0.45 | 0.95 |
| | Sn% | Sb% | W% | Ni% | Ceq% |
| | 0.13 | 0.192 | 0.183 | 0.24 | 0.42 |

To further investigate the thermal deformation behavior of materials under high temperature conditions, samples were taken along the direction of the continuous casting billet near the surface of the casting billet and subjected to hot tensile tests using a Gleeble 3500 thermal simulator. The hot tensile test used a cylindrical sample with a diameter of 10 mm. First, the sample was heated at a rate of 10 °C/s to 1320 °C and maintained for 4 min, and then cooled to the target temperature (600~1300 °C) at a cooling rate of 10 °C/s and maintained for 2 min. After that, the sample was stretched at a strain rate of $10^{-3}$ $s^{-1}$. When the sample fractured, an inert gas was used to rapidly cool the sample. A Zeiss Axio Scope A1 metallographic microscope, KEYENCE VHX5000 digital microscope, and Thermo-Fisher Apreo-S HiVac field emission scanning electron microscope were used to observe the cracks in the casting billet, the fracture surface and longitudinal section of the samples, and the distribution of crack elements. Additionally, the Shimadzu 8050 G electron probe microanalysis was used to improve the recognition of the elemental influence on the initiation of cracks in billets.

## 3. Results

### 3.1. Crack Morphology and Elemental Analysis of 09CrCuSb Continuous Casting Billet

Figure 1 depicts the cross-sectional morphology of cracks observed within continuous casting billets, characterized by two specific regions. In Figure 1a, the presence of a main crack extending along the plate-like ferrites toward the surface of the billets evidences its significance in the initial stages of crack formation, alongside cracks originating from inside the ferrite. In contrast, Figure 1b demonstrates a crack with apparent secondary cracks following the ferrite direction. The figure shows that the depth of the billet cracks exceeds 700 μm at the deepest point, and the austenite structure around the cracks is relatively coarse. Thin-film pearlite is distributed in a net-like manner along the original austenite structure, while the cracks exhibit characteristics of extension along the pearlite thin film. Inside the original austenite grains, the structure is characterized by small, interlocking acicular pearlite. Further analysis was conducted on the elements around the crack area in Figure 1b using energy spectrum analysis under a scanning electron microscope, as shown in Figure 2. It can be clearly observed that the elements Cr, Cu, and Sb are enriched along the interior of the crack.

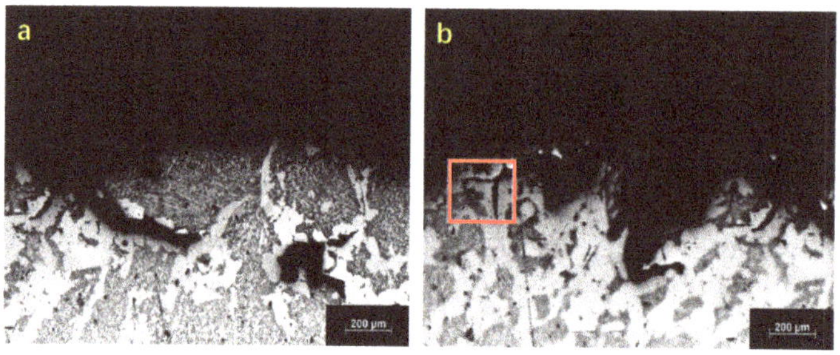


**Figure 1.** Microscopic morphology of cracks in continuous casting billets, regions 1 (**a**) and 2 (**b**).

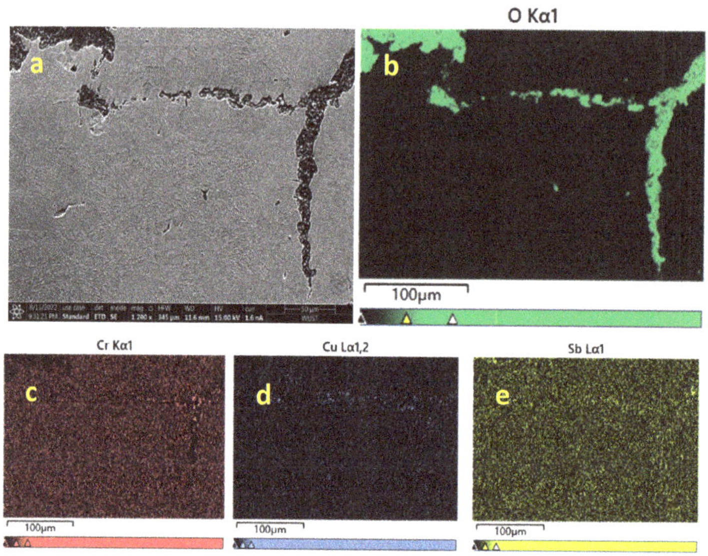


**Figure 2.** SEM morphology (**a**) and energy spectrum maps (**b–e**) of local surface cracks on continuous casting billets.

*3.2. High-Temperature Thermo-Plasticity of 09CrCuSb Steel*

Figure 3 shows the tensile properties and high-temperature strength and section shrinkage rate at different temperatures. The calculation formula for the section shrinkage rate of the tensile fracture is $A = (D_0{}^2 - D_i{}^2)/D_i{}^2$, where $D_0$ is the original diameter of the sample, and $D_i$ is the diameter of the tensile fracture at different temperatures. From the trend of strength change, as the insulation temperature increases, the overall tensile strength of the material tends to decrease, which is similar to the results reported in [19]. For high-temperature plasticity of the material, the section shrinkage rate of the tensile fracture gradually increases with the temperature rising in the range of 600–750 °C, and the section shrinkage rate is greater than 40% in this temperature range. Figure 4a–c show obvious necking features of the fracture, indicating that the plasticity of the material is better in this temperature range [20]. Combined with the fracture morphology under SEM in Figure 5a–c, the fracture presents obvious ductile dimples at 600 °C, and some inclusions can be observed at the bottom of some dimples. As the temperature further increases, the depth of the dimples becomes shallower and the internal pattern becomes fragmented, and there are only a few ductile dimple features, indicating that the plasticity of the material decreases with the temperature increase.

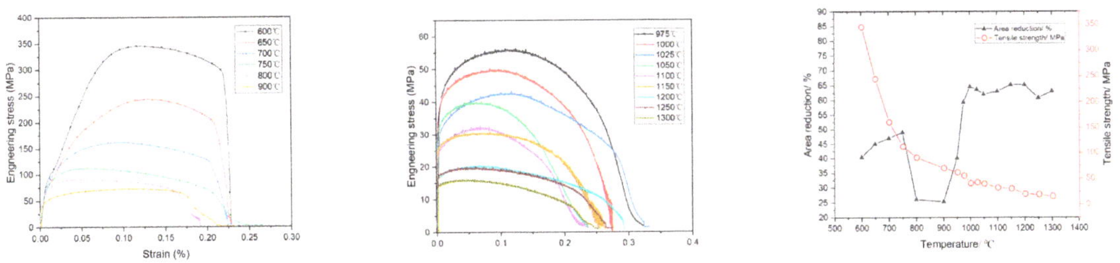


(**a**) (**b**) (**c**)

**Figure 3.** Thermal tensile stress–strain curves (**a**,**b**) at different temperatures, changes in thermal tensile strength and reduction of area (**c**).

When the hot tensile test is conducted in the temperature range of 800–900 °C, the section shrinkage rates at the two temperatures are 26.0% and 25.4%, respectively, and the section shrinkage rate of the samples decreases significantly, indicating that the plasticity of the material is significantly reduced. From the macroscopic morphology of the fracture (such as Figure 4d–f), when the tensile temperature is 800 °C, the hot tensile fracture has a flat step surface but no obvious necking features, and the section shrinkage rate of the samples is lower and the plasticity inferior. Further observation of the fracture morphology under scanning electron microscopy (Figure 5d) shows typical brittle cleavage features on the fracture. Unlike the results obtained by Hu et al. [10,19], the plasticity of the 09CrCuSb steel under the new component system is in a trough zone between 800 and 900 °C, which is the third brittle temperature range of the 09CrCuSb steel.

When the hot tensile temperature is above 1000 °C, the section shrinkage rate increases significantly to over 60%, the macroscopic fracture shows necking features (such as Figure 4f–h), and the section is relatively flat. The SEM morphology of the fracture shows shallower ductile dimples, as shown in Figure 5c. As the temperature further increases, there are many high-temperature molten beads on the fracture that are difficult to further analyze, but the necking features become more obvious, as shown in Figure 4d–i. When the temperature is above 1100 °C, the section shrinkage rate further increases and the necking features become more obvious, showing good high-temperature plasticity.

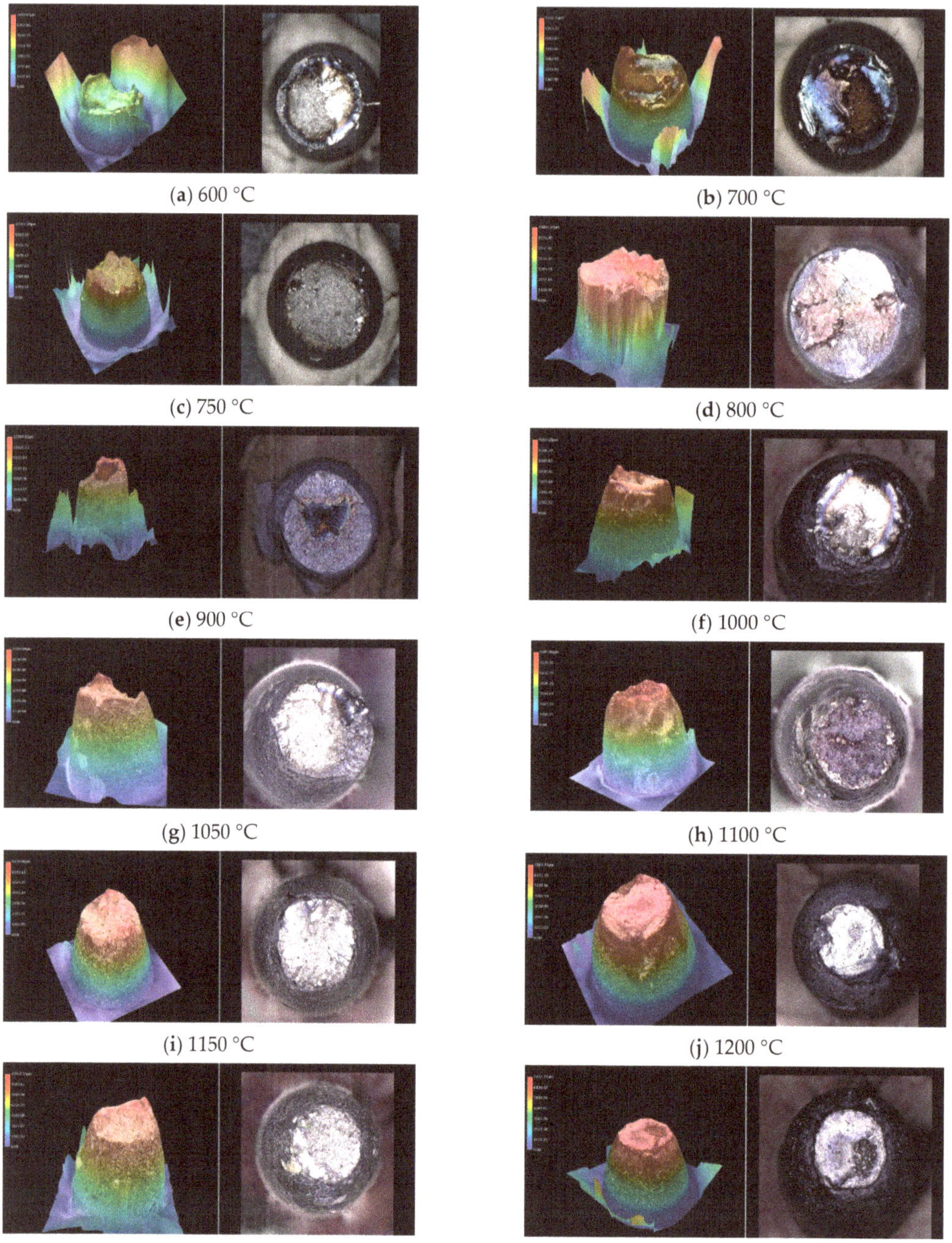

(a) 600 °C (b) 700 °C

(c) 750 °C (d) 800 °C

(e) 900 °C (f) 1000 °C

(g) 1050 °C (h) 1100 °C

(i) 1150 °C (j) 1200 °C

(k) 1250 °C (l) 1300 °C

**Figure 4.** Macromorphology of hot stretching fractures at different temperatures.

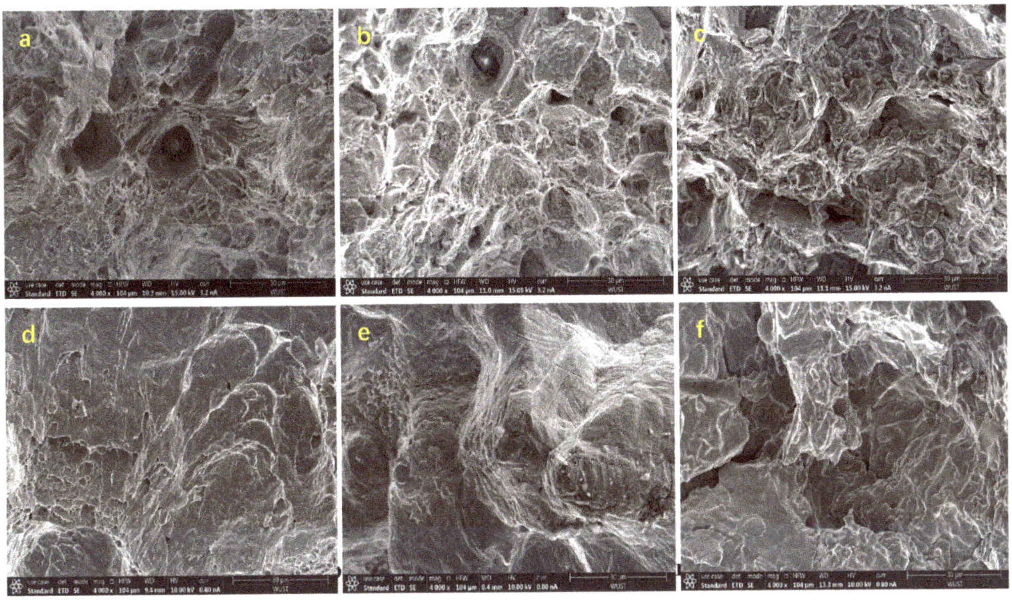


**Figure 5.** SEM morphology of hot tensile fracture at different temperatures: 600 °C (**a**), 700 °C (**b**), 750 °C (**c**), 800 °C (**d**), 900 °C (**e**), 1000 °C (**f**).

When the stress on the material is between the yield point strength and the tensile strength, the specimen will undergo uniform plastic deformation. At this time, a small deformation amount will cause stress concentration, causing the stress on the material to quickly increase to the ultimate strength and cause fracture. By calculating the plastic modulus of the deformation stage, it can be used as an evaluation index of material plastic strengthening. For high-temperature tensile cast specimens, the larger the plastic modulus, the worse the plasticity, that is, the more likely it is to break at a certain temperature [21]. Based on the Hollomon formula, the plastic modulus of the material satisfies the following relationship [22,23].

$$\sigma = H\varepsilon^n \tag{1}$$

Among them, ε is the true strain of the material, $\sigma$ is the true stress, and $H$ is the plastic modulus of the material. Since the strain rate in this experiment can be considered constant, taking the logarithm of Equation (1) on both sides and differentiating it, by calculating the instantaneous slope value of the true stress–strain curve in the logarithmic coordinate system, we can further obtain the hardening index $n$ of the material [22,23]:

$$n = \frac{dlg\sigma}{dlg\varepsilon} \tag{2}$$

The relationship between the plastic modulus $H$ and the temperature of the material, as shown in Figure 6, can be obtained by processing the true stress–strain curves under different temperatures. The $n$ values in the fitting process are distributed in the range of 0.05 to 0.5, indicating a good fitting result [23]. From the variation of the plastic modulus obtained by the fitting with the temperature, the overall trend of the plastic modulus is decreasing. At temperatures of 600–750 °C, the plastic modulus value is relatively large but the change relatively flat, with the highest value appearing at 700 °C. When the temperature is above 750 °C, the plastic modulus value rapidly decreases, and the change slows down after 900 °C. The highest value appears again at 975 °C, and finally slowly decreases with the increase of temperature. From the results of the high-temperature plastic modulus, it can be seen that the plasticity of 09CrCuSb steel is poor around 700 °C, so it should be avoided as much as possible to have significant deformation at 700 °C during hot rolling. When the temperature is above 800 °C, the plastic deformation ability of 09CrCuSb steel also decreases significantly, so in the continuous casting process, the two-water control process should also avoid the billet staying at a temperature of 800–975 °C for a long time [21].

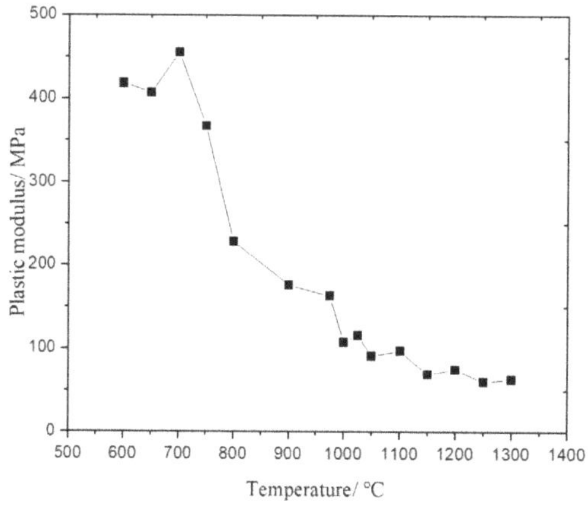


**Figure 6.** Change of plastic modulus (dotted line) of 09CrCuSb steel with temperature.

### *3.3. Microstructure Deviation and Elements Segregation near Cracks in 09CrCuSb Steel during the High-Temperature Tensile Process*

The organizational characteristics of the longitudinal section of the fracture after hot tensile testing of 09CrCuSb steel at different temperatures are shown in Figure 7a–i. From Figure 7a, it can be seen that when the tensile temperature is 600 °C, the structure near the fracture is composed mainly of ferrite and pearlite. Observing the secondary cracks near the fracture, it can be seen that the cracks extend along the original austenite grains. Comparing Figure 7a–c, it can be seen that as the tensile temperature increases, both the ferrite and pearlite structures undergo obvious coarsening, which is consistent with the grain isothermal growth model proposed by E. Anelli et al. [24]. When the temperature reaches 800 °C, as shown in Figure 7d, the fracture and secondary cracks extend mainly along the original austenite grain boundary, and there is an obvious film-like proeutectoid ferrite structure near the original austenite grain boundary, which is consistent with the observed organizational characteristics near the casting crack, and the structure near the center of the austenite grain is displayed mainly as granular bainite and acicular ferrite. When the hot tensile temperature is 900 °C, the longitudinal section of the fracture and secondary cracks also show the characteristic of extension along the proeutectoid ferrite, and the structure near the center is mainly acicular ferrite, as shown in Figure 7e. Comparing Figure 7d–f, as the temperature increases within 800–1000 °C, the grain size of the ferrite structure around the original austenite grain boundary tends to increase, and the thickness of the film-like proeutectoid ferrite at the austenite grain boundary also gradually increases. When the hot tensile temperature is greater than 1050 °C (as in Figure 7g–i), the film-like proeutectoid ferrite on the longitudinal section of the fracture disappears, and the internal organization near the fracture is composed mainly of bainite. When the hot tensile temperature is 1300 °C, the fracture surface becomes flat and displays the characteristics of Type I high-temperature brittle fracture.

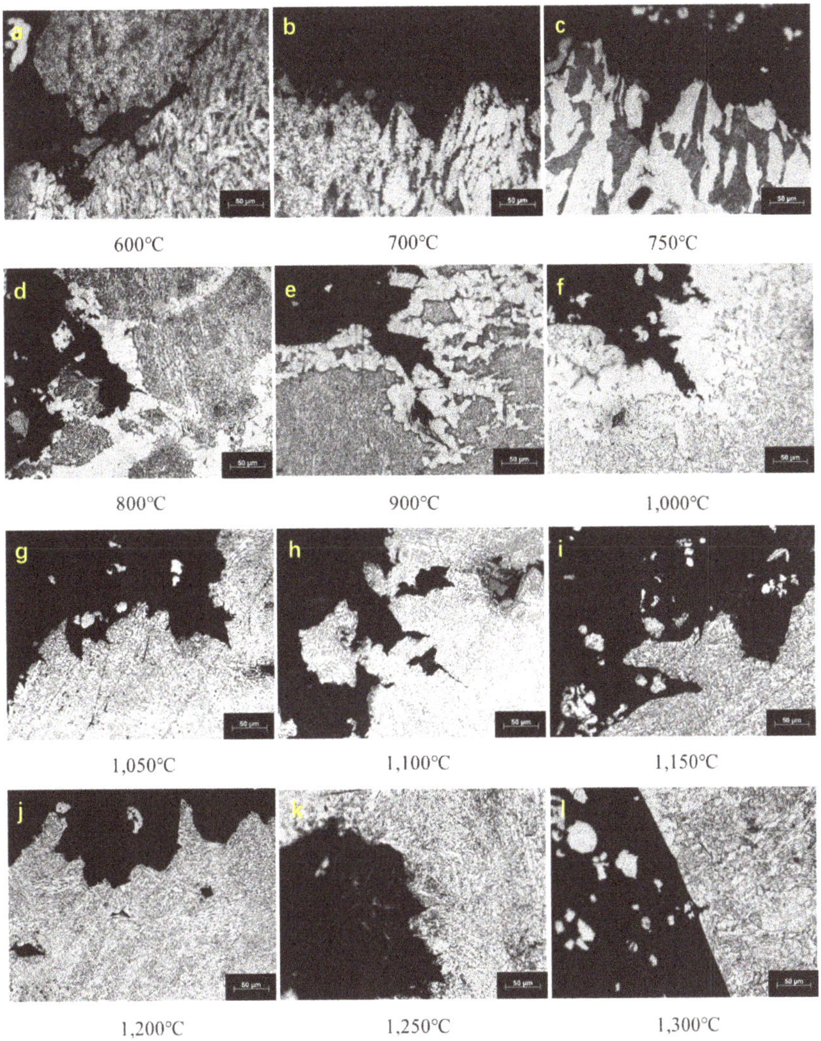


**Figure 7.** Metallographic structure of hot stretching fracture section at different temperatures.

The EPMA analysis of the crack near the tested area under 800 °C thermal tensile conditions are depicted in Figure 8. The elemental map indicates a higher content of carbon in the bainites and at the grain boundaries of proeutectoid ferrites. Furthermore, Cu exhibits clear segregation at the ferrite triple junctions as red circled in Figure 8, whereas there is no noticeable segregation behavior for Cr. Zhou et al. [14] have similarly concluded that elevated temperatures during continuous casting exhibit a high level of Cu solubility, with a thermal threshold exceeding 900 °C. Yet, the low diffusion coefficient of Cu within Fe presents a substantial obstacle for the material to infiltrate the steel matrix, particularly where the grain boundary has experienced Cu buildup.

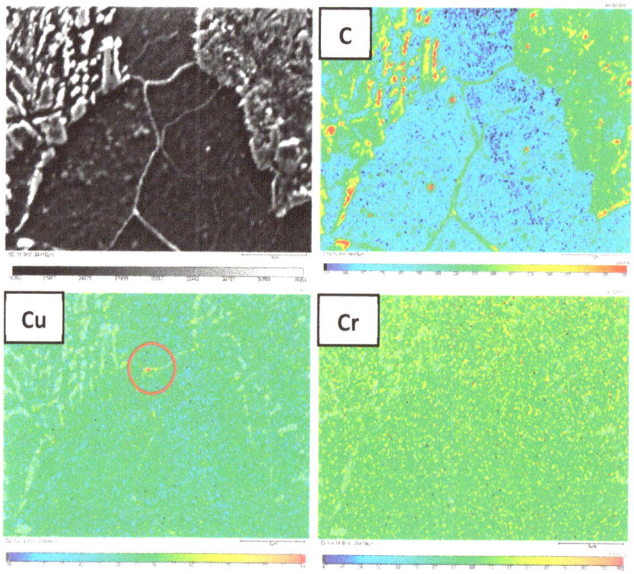


**Figure 8.** EPMA analysis on the crack face adjacent area after fracture under 800 °C.

## 4. Discussion

From Figures 1 and 7, it can be observed that there are obvious film-like proeutectoid ferrite grain boundaries near the crack and the 800–1000 °C tensile fracture surface of the billet, and secondary cracks also mainly occur and propagate in the proeutectoid ferrite grain boundaries distributed along the austenite grain boundaries. According to the classification of allotriomorphic ferrite by Yang et al. [25], both sides of the allotriomorphic ferrite near the crack in the billet have uneven and unclear boundary contours and belong to the GF-II type allotriomorphic ferrite, which is formed by the nucleation of original austenite grain boundaries and growth inward. Li et al. [26] pointed out that one of the essential influences of austenite deformation on the nucleation and incubation period of proeutectoid ferrite is that the deformation intensifies the disorder of the atomic arrangement on the austenite grain boundary, causing the increase of grain boundary energy and promoting the nucleation of proeutectoid ferrite. Thus, during the process of casting, the high temperature provides the formation condition of austenite, and the casting forming provides the intensified disorder of the atomic arrangement. The plate-like proeutectoid ferrites is easy to nucleate along the austenite boundaries.

It should be noted that compared with the old version of the energy standard NB/T47019-2011 [27], the newly issued NB/T47019-2021 standard has increased the content of proeutectoid-forming elements such as Cr, Mo, Si, Nb, Al, and Ti [28,29], which is represented by the Ae3 temperature increasing from 855 °C to 880 °C on the phase diagram, as calculated by the Thermal-Calc software in Figure 9. Ae3 temperature represents the beginning temperature of the thermodynamic austenite–ferrite transformation [30–33]. It can be seen that the increase of alloy elements means that the transformation time of proeutectoid ferrite during the cooling process is advanced due to the increase of Ae3 temperature [33,34]. It also means that during the cooling process, the formation of ferrite will happen before it would in steel [35]. Therefore, when the slab enters the straightening machine during continuous casting, the surface temperature due to the effect of reheating should return to above 800 °C. The expansion of the two-phase region leads to the extension of the insulation time of the billet in the two-phase region, and proeutectoid ferrite precipitation is also promoted when the grain boundary satisfies certain thermodynamic conditions. Given the crucial role of proeutectoid ferrite formation, it is important to give due consideration to carbon content.

The diffusion of carbon atoms into the inner part of the austenite results in the duplication of the original austenite grain boundary, which is responsible for the eutectoid ferrite as well [11,36]. At the same time, some scholars have pointed out that in steels with high microalloyed element content, carbon and nitrogen will precipitate and aggregate at the grain boundaries [37–39], further deteriorating the high-temperature mechanical properties of the III brittle zone and causing transverse cracks and corner cracks in the billet.

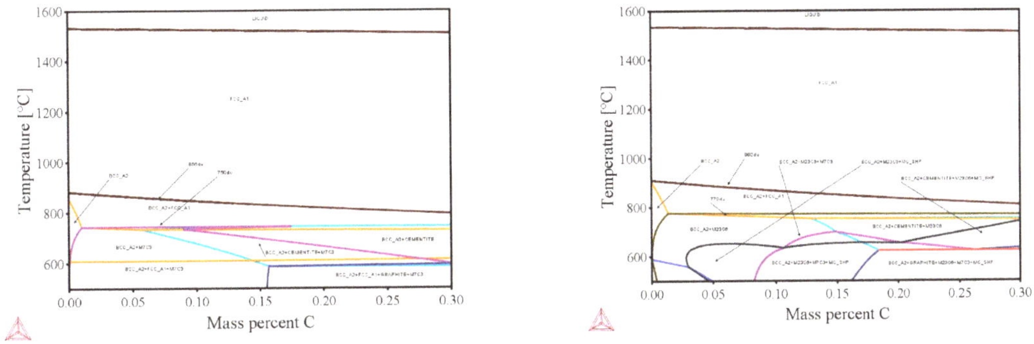


(**a**) (**b**)

**Figure 9.** Equilibrium phase diagram calculated based on the old composition frame (**a**) and the current (**b**).

Moreover, due to the typical condition of non-grain-boundary equilibrium segregation during the cooling process of the billet, the segregation of elements at grain boundaries is also an important factor leading to the crack in the billet. As depicted in Figure 2, there is segregation of the elements Cr, Cu, and Sb. Anthony pointed out that non-equilibrium segregation is formed by the combination of solute atoms, vacancies, and solute atoms and high-temperature vacancies under certain conditions, and then a large number of complex bodies move to the grain boundaries during the cooling process [40,41]. The temperature difference of cooling determines the maximum degree of non-equilibrium segregation concentration. Usually, the greater the temperature difference of cooling process, the higher the segregation concentration, which is called the thermal gradient effect [42,43]. From the distribution of elements around the crack in the billet, it can be seen that Cr, Cu, and Sb are enriched near the crack, indicating that these easily segregated and high-content elements also play a significant role in crack initiation and propagation. For the grain boundary segregation of Cr, Kazuma et al. [8] reported that Cr shows high segregation energy at sites where the coordination number of the grain boundary is high and the grain size is large. Zhou et al. [14] summarized the work of G. Salje [44] and implied that when the temperature exceeds 925 °C, although the solubility of Cu in Fe increases to over 4%, the corresponding diffusion coefficient is very low. In continuous casting, the billet is heated above 900 °C, hindering the Cu in the grain boundary from diffusing into the steel matrix to create a copper-rich phase on the billet surface. There is a greater likelihood of crack formation along the grain boundary, notably for billets subject to bending and straightening deformation. In contrast, within the temperature range of 800–900 °C, there exists a region where the diffusivity of Cu and the solubility are both high. Thus, the segregation of Cu is severe and detrimental to the casting billets. As the typical low melting point element, Sb also has high tendency to segregate at the grain boundaries [45]. The segregation of Cu and Sb at grain boundaries of steel could weaken the grain boundaries, and the weakened grain boundaries were cracking under thermal stress [46].

In summary, for 09CrCuSb steel in the new composition frame, in order to reduce the sensitivity of billet cracking during continuous casting, the secondary cooling water control process should be adjusted to avoid the billet staying at a temperature range of 800–900 °C for a long time.

## 5. Conclusions

(1) The cracking observed on the surface of 09CrCuSb steel castings can be attributed to two primary factors. First, the modification in alloy composition results in an increase in the Ae3 temperature (from 855 °C to 880 °C), leading to the precipitation of a thin, film-like coexisting ferrite along the grain boundaries. Second, alloying elements have a higher propensity to experience non-equilibrium segregation at grain boundaries, which weakens the austenite grain boundary strength. This feature eventually leads to surface embrittlement and the formation of cracks on the castings.

(2) The 09CrCuSb steel has poor plasticity at 700 °C, making it unsuitable for hot rolling in the pipe-making process. Between 800 and 900 °C, the steel enters the third brittle temperature range. To prevent damage, castings need to avoid staying in this temperature range for too long during the two-phase cooling process of continuous casting.

(3) The stress–strain curve of 09CrCuSb steel at various test temperatures shows a gradual decrease in its plastic modulus with increasing temperature. The material exhibits good plastic deformation between 600 and 650 °C. When the temperature exceeds 1000 °C, the plastic modulus is low (<80 MPa), indicating that the castings can withstand large uniform plastic deformation at high temperatures.

**Author Contributions:** Conceptualization, Z.P.; methodology, T.M. and Z.P.; validation, Z.P., T.M. and Y.Y.; formal analysis, Z.P.; investigation, Z.P., Y.Y. and J.Z.; resources, J.Z. and Y.Y.; data curation, J.Z. and Y.Y.; writing—original draft preparation, Z.P. and T.M.; writing—review and editing, J.Z.; supervision, L.W.; project administration, J.Z. and L.W.; funding acquisition, Z.P. All authors have read and agreed to the published version of the manuscript.

**Funding:** This work is funded by National Natural Science Foundations of China (52201084).

**Institutional Review Board Statement:** Not applicable.

**Informed Consent Statement:** Informed consent was obtained from all subjects involved in the study.

**Data Availability Statement:** Because data is unavailable due to privacy, where no new data were created.

**Acknowledgments:** Postdoctoral Station of Metallurgical Engineering of Wuhan University of Science and Technology, Postdoctoral Workstation of Zhejiang Jincheng New Material Co., Ltd. This work also thanks the Analysis and Testing Centre of Wuhan University of Science and Technology.

**Conflicts of Interest:** The authors (Zhixian Peng, Tao Mei, Jian Zheng, Yuan Yuan, Liwang Wang) declare that we have no financial and personal relationships with other people or organizations that could inappropriately have influenced our work, there is no professional or other personal interest of any nature or kind in any product, service and/or company that could be construed as influencing the position presented in or review of this manuscript. The authors declare no conflict of interest.

## References

1. Cai, H. Application of 09CrCuSb Steel in Boiler Manufacturing. *Ind. Boil.* **2005**, *5*, 22–25.
2. Ling, L.; Xu, X.; Fu, H. Welding Practice of 09CrCuSb Sulfuric Acid Dew Point Corrosion Resistant Steel. *Medium Nitrogen Fertil.* **2011**, *2*.
3. *NB/T47019-2021*; Technical Conditions for Ordering Pipes for Boilers and Heat Exchangers. National Energy Administration of China: Beijing, China, 2021.
4. Zuo, W.; Zhang, X.; Li, Y. Review of flue gas acid dew-point and related low temperature corrosion. *J. Energy Inst.* **2020**, *93*, 1666–1677. [CrossRef]
5. Zeng, Y.; Li, K.; Hughes, R.; Luo, J.-L. Corrosion mechanisms and materials selection for the construction of flue gas component in advanced heat and power systems. *Ind. Eng. Chem. Res.* **2017**, *56*, 14141–14154. [CrossRef]
6. Chen, Z.; Wang, Q.; Cai, Y.; Wang, Z. Analysis on the Formation Mechanism of Surface Cracks on 09CrCuSb Billets. *Mod. Transp. Metall. Mater.* **2021**, *1*, 83–87.
7. Yu, H.; Wang, Y.; Zhang, J.; Yang, S.; Liu, X. Review on the Causes of Surface Cracks in Continuous Casting Billets and Their Evolutionary Behavior during Rolling Process. *Forg. Stamp. Technol.* **2010**, *2*, 1–5.

8. Ito, K.; Tanaka, Y.; Sawada, H. Application of Grain Boundary Segregation Prediction Using a Nano-Polycrystalline Grain Boundary Model to Transition Metal Solute Elements: Prediction of Grain Boundary Segregation of Mn and Cr in bcc-Fe Polycrystals. *Mater. Trans.* **2022**, *63*, 269–277. [CrossRef]
9. Han, J.P.; Li, Y.; Jiang, Z.H.; Yang, Y.C.; Wang, X.X.; Wang, L.; Li, K.T. Summary of the Function of Sn in Iron and Steel. In *Advanced Materials Research*; Trans Tech Publications Ltd.: Bäch, Switzerland, 2013; pp. 406–411.
10. Hu, H.; Zhu, L.; Sun, L.; Zhou, J. Mechanism and Preventive Measures of Corner Cracks in 09CrCuSb Steel Continuous Casting Billets. *Shanghai Met.* **2020**, *42*, 7.
11. Chen, Y.; Ji, C.; Zhu, M. Prediction Model of the Proeutectoid Ferrite Growth of a Continuous Casting Slab. *Metall. Mater. Trans.* **2023**, *54*, 1101–1115. [CrossRef]
12. Wei, F.; Cheng, B.; Chew, L.T.; Lee, J.J.; Cheong, K.H.; Wu, J.; Zhu, Q.; Tan, C.C. Grain distribution characteristics and effect of diverse size distribution on the Hall–Petch relationship for additively manufactured metal alloys. *J. Mater. Res. Technol.* **2022**, *20*, 4130–4136. [CrossRef]
13. Cao, J.; Dong, J.; Saglik, K.; Zhang, D.; Solco, S.F.D.; You, I.J.W.J.; Liu, H.; Zhu, Q.; Xu, J.; Wu, J. Non-equilibrium strategy for enhancing thermoelectric properties and improving stability of AgSbTe2. *Nano Energy* **2023**, *107*, 108118. [CrossRef]
14. Zhou, J.; Zhu, L.; Sun, L.; Wang, B.; Xiao, P. Analysis of the Formation Mechanism of Surface Cracks of Continuous Casting Slabs Caused by Mold Wear. *Processes* **2022**, *10*, 797. [CrossRef]
15. Al-Asad, M.; Alam, M.N.; Tunç, C.; Sarker, M. Heat Transport Exploration of Free Convection Flow inside Enclosure Having Vertical Wavy Walls. *J. Appl. Comput. Mech.* **2021**, *7*, 520–527.
16. Akinshilo, A.; Davodi, A.; Rezazadeh, H.; Sobamowo, G.; Tunç, C. Heat Transfer and Flow of Mhd Micropolarnanofluid through the Porous Walls, Magnetic Fields and Thermal Radiaton. *Palest. J. Math.* **2022**, *11*, 604–616.
17. Lu, T. *Analysis on the Causes of Surface Cracks in 09CrCuSb Billets and Improvement Measures*; Anhui University of Technology: Ma'anshan, China, 2016.
18. Huang, X. *The Effect of Grain Boundary Segregation of Residual Elements As, Sn, Sb on the Thermoplasticity of Continuous Casting Steel*; Wuhan University of Science and Technology: Wuhan, China, 2006.
19. Hu, H.; Zhu, L.; Sun, L.; Zhou, J. High Temperature Mechanical Properties of 09CrCuSb Steel Continuous Casting Billets. *Iron Steel Vanadium Titan.* **2020**, *41*, 132–136.
20. Mintz, B. The influence of composition on the hot ductility of steels and to the problem of transverse cracking. *ISIJ Int.* **1999**, *39*, 833–855. [CrossRef]
21. Chen, D.; Huang, L.; Wang, Y.; Wen, L.; Feng, K.; Liu, Q.; Han, Z. Experimental Study on High Temperature Plastic Modulus of Q345 Steel. In Proceedings of the 2006 National Metallurgical Physical Chemistry Academic Conference of China, Jinan, China, 29–30 December 2006.
22. Zener, C.; Hollomon, J.H. Effect of strain rate upon plastic flow of steel. *J. Appl. Phys.* **1944**, *15*, 22–32. [CrossRef]
23. Meyers, M.A.; Chawla, K.K. *Mechanical Behavior of Materials*; Cambridge University Press: Cambridge, UK, 2008.
24. Anelli, E. Application of mathematical modelling to hot rolling and controlled cooling of wire rods and bars. *ISIJ Int.* **1992**, *32*, 440–449. [CrossRef]
25. Yang, Z.; Wang, K.; Sun, X.; Da, G. Morphology of grain boundary ferrite and grain boundary cracks in niobium titanium microalloyed steel billets. *Iron Steel* **2018**, *53*, 10.
26. Li, J.; Sun, F. The Effect of Austenitic Deformation on the Incubation Period of Isothermal Formation of Pre eutectoid Ferrite. *Acta Metall. Sin.* **1990**, *4*, 161–164.
27. *NB/T47019-2011*; Technical Conditions for Ordering Pipes for Boilers and Heat Exchangers. National Energy Administration of China: Beijing, China, 2011.
28. Liu, M.; Li, Y.; Cui, Z.; Yang, Q. High ductility of spray formed low density TRIP steel with the improvement of δ-ferrite matrix. *Mater. Charact.* **2019**, *156*, 109828. [CrossRef]
29. Shao, Y.; Liu, C.; Yan, Z.; Li, H.; Liu, Y. Formation mechanism and control methods of acicular ferrite in HSLA steels: A review. *J. Mater. Sci. Technol.* **2018**, *34*, 737–744. [CrossRef]
30. Equihua-Guillén, F.; Salinas-Rodriguez, A. Role of the austenite-ferrite transformation start temperature on the high-temperature ductility of electrical steels. *J. Mater. Eng. Perform.* **2011**, *20*, 102–107. [CrossRef]
31. Ghosh, C.; Aranas, C., Jr.; Jonas, J.J. Dynamic transformation of deformed austenite at temperatures above the Ae3. *Prog. Mater. Sci.* **2016**, *82*, 151–233. [CrossRef]
32. Toloui, M.; Militzer, M. Phase field modeling of the simultaneous formation of bainite and ferrite in TRIP steel. *Acta Mater.* **2018**, *144*, 786–800. [CrossRef]
33. Maubane, D.R.N.; Mostert, R.J.; Banks, K.M. Re-Austenitisation of Thin Ferrite Films in C–Mn Steels during Thermal Rebound at Continuously Cast Slab Corner Surfaces. *Metals* **2022**, *12*, 2155. [CrossRef]
34. Aaronson, H.; Reynolds, W.; Purdy, G. Coupled-solute drag effects on ferrite formation in Fe-CX systems. *Metall. Mater. Trans. A* **2004**, *35*, 1187–1210. [CrossRef]
35. Kwok, T.; Dye, D. A review of the processing, microstructure and property relationships in medium Mn steels. *Int. Mater. Rev.* **2023**, 1–37. [CrossRef]
36. Yin, J.; Hillert, M.; Borgenstam, A. Morphology of proeutectoid ferrite. *Metall. Mater. Trans. A* **2017**, *48*, 1425–1443. [CrossRef]

37. Li, F.; Tian, J.; Li, H.; Deineko, L.; Jiang, Z. Simultaneously Enhancing Strength, Ductility and Corrosion Resistance of a Martensitic Stainless Steel via Substituting Carbon by Nitrogen. *Acta Metall. Sin. (Engl. Lett.)* **2023**, *36*, 705–716. [CrossRef]
38. Liu, F.; Kang, C.; Qian, R.; Jiang, Z.; Geng, X.; Li, H. Effect of Tempering Temperature on Microstructure and Properties of a New Type of Nitrogen—Containing Hot-Work Die Steel 3Cr7Mo2NiSiVN. *Steel Res. Int.* **2022**, *93*, 2200013. [CrossRef]
39. Yang, Y.; Jia, X.; Ma, Y.; Wang, P.; Zhu, F.; Yang, H.; Wang, C.; Wang, S. Effect of Nb on microstructure and mechanical properties between base metal and high heat input coarse-grain HAZ in a Ti-deoxidized low carbon high strength steel. *J. Mater. Res. Technol.* **2022**, *18*, 2399–2412. [CrossRef]
40. Anthony, T.R. Solute segregation in vacancy gradients generated by sintering and temperature changes. *Acta Metall.* **1969**, *17*, 603–609. [CrossRef]
41. Yang, S.; Ma, J.; Chen, C.; Zhang, C.; Ren, J.; Jiang, Z.; Fan, G.; Han, P. Effects of B and Ce Grain Boundary Segregation on Precipitates in Super Austenitic Stainless Steel. *Metals* **2023**, *13*, 326. [CrossRef]
42. Xu, T. Non-equilibrium grain-boundary segregation kinetics. *J. Mater. Sci.* **1987**, *22*, 337–345. [CrossRef]
43. Li, Q.; Li, L.; Zheng, L.; Liu, F.; Guo, J.; Shanglin, Y. Non-equilibrium grain boundary segregation kinetics of phosphorus in 12Cr1MoV steel at different solution temperatures. *J. Mater. Sci.* **2004**, *39*, 6551–6554.
44. Salje, G.; Feller-Kniepmeier, M. The diffusion and solubility of copper in iron. *J. Appl. Phys.* **1977**, *48*, 1833–1839. [CrossRef]
45. Babapour, A.; Hosseinipour, S.J.; Jamaati, R.; Abbasi, M. Effect of Sb and Initial Annealing on the Microstructure, Texture, and Magnetic Behavior of Low Silicon Steel Produced by Single-Roll Drive Rolling. *Met. Mater. Int.* **2022**, *29*, 1815–1824. [CrossRef]
46. Chen, C.; Zhou, J.; Yu, J.; Ju, J.; Zhang, Y.; Wang, J.; Fan, B. Interfacial microstructures and infiltrated cracks in tin bronze/steel bimetallic materials fabricated by arc cladding. *J. Mater. Sci.* **2023**, *58*, 4679–4693. [CrossRef]

*metals*

MDPI

*Article*

# Solidification Kinetics of an Al-Ce Alloy with Additions of Ni and Mn

**Jordan Roger Kozakevich [1,*], Joshua Stroh [1], Dimitry Sediako [1] and David Weiss [2]**

[1] High-Performance Powertrain Materials Laboratory, University of British Columbia—Okanagan, 1137 Alumni Ave., Kelowna, BC V1V 1V7, Canada; joshua.stroh@ubc.ca (J.S.); dimitry.sediako@ubc.ca (D.S.)
[2] Eck Industries, 1602 N 8th St., Manitowoc, WI 54220, USA; david.weiss@eckindustries.com
* Correspondence: jordan.kozakevich@ubc.ca

**Abstract:** Heat-treated aluminum–silicon (Al-Si)-based alloys have dominated the cast lightweight alloy industry for several decades. However, in the last decade, Al-Ce-based alloys have shown promise in replacing Al-Si alloys as they remove the need for costly heat treatments. As the properties of Al-Ce alloys depend on the as-cast microstructure, it is important to characterize the solidification kinetics of these alloys. Therefore, this study focused on characterizing the solidification of an Al-Ce alloy with additions of Ni and Mn (nominal composition Al-12.37Ce-3.26Ni-0.94Mn-0.12Fe in weight percent). The alloy was cast in a wedge mold configuration, resulting in cooling rates between 0.18 and 14.27 °C/s. Scanning electron microscopy (SEM) coupled with the energy dispersive x-ray spectroscopy (EDS) and differential scanning calorimetry (DSC) techniques characterized the evolution rate of solid phases. The SEM/EDS data revealed that an $Al_{10}CeMn_2$ phase was present at higher cooling rates. At lower cooling rates, near the center of the casting, a primary $Al_{23}Ce_4Ni_6$ phase was more present. It was observed that up to 2.6 atomic percent (at.%) of Mn was dissolved in this primary $Al_{23}Ce_4Ni_6$ phase, thereby removing a large portion of the available Mn for forming the $Al_{10}CeMn_2$ phase. DSC analysis showed differences in the samples' liquidus temperatures, which indicated compositional variations. Inductively coupled plasma–atomic emission spectroscopy (ICP-OES) and Scheil solidification simulations correlated the compositional differences with phase formation, which agreed with the SEM and DSC results. This experiment provides insight into novel Al-Ce-Ni-Mn alloys and where their potential lies in industrial applications.

**Keywords:** solidification kinetics; aluminum; cerium

## 1. Introduction

With the growth of environmentally friendly technology, there is a high demand dependency on rare earth (RE) element mining. RE elements, such as neodymium, praseodymium, dysprosium, and samarium, are used for high-performance magnets in applications such as electric cars and wind turbines [1–3]. However, the desirable RE elements typically comprise approximately 25% of the total material mined. The remaining 75% is discarded as a by-product and has little industrial relevance [4]. Of the RE elements discarded, cerium (Ce) is the most abundant, at 38% [5]. The usefulness of Ce, however, has been researched throughout the 20th century concerning alloying it with aluminum (Al) to help with issues of insufficient thermal stability. For example, Belov et al. [6], in 1999, conducted elevated-temperature experiments on several Al alloys with additions of Ce and nickel (Ni). Targeting 350 °C, Belov and his colleagues found that an Al-12Ce-5Ni (wt.%) alloy exhibited a 75% increase ultimate tensile strength (UTS) over a conventional Al-Si alloy (339) used at the time. However, this research was not continued due to the high cost of Ni and the lack of a Ce supply chain. That is, until the demand for high-performance magnets experienced a dramatic increase, resulting in an abundance of available Ce.

This abundance has captured the interest of researchers investigating how to improve the efficiency of internal combustion engines by creating thermally stable lightweight alloys for next-generation automotive powertrains [7–11]. A clear emergence from all of this research is the potential for the Al-Ce-based alloying system to become a fundamental pillar of high-performance Al alloys that can operate in applications that require excellent thermal stability above 200 °C [12–16]. Extensive research has explored the fundamentals of the Al-Ce binary system to understand the platform from which a new alloying system can be structured. The solidification characteristics and phase analysis of hypoeutectic [17,18], eutectic [13,18–20], and hypereutectic [17–19,21] binary alloy compositions revealed that the $Al_{11}Ce_3$ phase has excellent castability and thermal stability up to 500 °C. In a hypereutectic composition, primary $Al_{11}Ce_3$ formation occurs. If the content of Ce exceeds 16 wt.%, the primary phase begins to crack because of the large coefficient of the thermal expansion discrepancy [7]. Therefore, most studies have focused on the eutectic composition to avoid any possible decrease in mechanical properties. These eutectic alloys precipitate a lamellae Al-$Al_{11}Ce_3$ phase that can retain up to 80% of its hardness when exposed to 500 °C for 168 h [13,20,22]. For comparison, the Al-Si eutectic composition only retains ~50% of its hardness when exposed to the same test conditions [20]. This evidence justifies building an elevated-temperature Al alloy system based on the Al-Ce binary system. The current major drawback of Al-Ce systems is that the lamellae Al-$Al_{11}Ce_3$ eutectic phase does not provide sufficient precipitate-strengthening benefits for next-generation powertrain applications.

This lack of strength of the Al-Ce eutectic phase brings relevance back to the research conducted by Belov et al. [6], where it was found that the addition of Ni enhanced the strength of these alloys. Although the testing parameters in the 1999 study were not rigorous enough to meet today's automotive industry demands, they provide insight into how minor alloy elements can be used to add strength to the Al-Ce system. Sims et al. [21] drew inspiration from Belov and his colleagues and investigated two higher-order Al-Ce-based alloys with magnesium (Mg) additions. The alloys under investigation were Al-12Ce-0.4Mg and Al-12Ce-4Si-0.4Mg (wt.%). It was concluded that the Al-Ce-Mg alloy in an as-cast state exhibited a UTS and yield strength (YS) of 200.6 and 78.6 MPa, respectively, while the Al-Ce-Si-Mg alloy exhibited values of 252.3 and 128.2 MPa, respectively, when heat treated to a T6 temper. These alloys were only tested at ambient temperature, but showed comparable strength to alloys such as T6 A356. Weiss et al. [23] took it a step further and tested two Al-Ce-Mg alloys, each with 8 wt.% Ce, but differing in Mg contents (7 vs. 10 wt.%). Both alloys were tested at ambient temperature and 260 °C. The alloy containing 7 wt.% magnesium exhibited a UTS and YS of 195 and 151 MPa, respectively. The alloy with 10 wt.% magnesium exhibited a UTS and YS of 227 and 186 MPa, respectively. When tested at 260 °C, the 7 wt.% Mg alloy retained 69% of its UTS and 80% of its YS. The alloy containing a larger amount of Mg retained 60% of its UTS and 70% of its YS. It is worth noting that these alloys were not conditioned (i.e., exposed to 260 °C for an extended period) before testing; therefore, it is expected that the alloys will perform slightly worse than expected if exposed to powertrain conditions for an extended period. It was also found in the study by Weiss et al. that the alloys were non-responsive to heat treatments, aside from homogenization having a positive effect on alloys with high Mg contents. As these Al-Ce-based alloys are typically not heat-treatable, they depend heavily on their solidification kinetics for their respective mechanical properties. The previous studies cited here do not account for the alloys' solidification kinetics, but they significantly impact the mechanical properties of the alloys. This solidification rate-dependency of Al-Ce alloys removes the time and resource-consuming heat treatments from the production process. Nguyen et al. [24] and Salonitis et al. [25] predicted that removing necessary heat treatments from Al alloy production alone could save 425,000–675,000 MWh annually. Along with the decreased production cost, decreasing this amount of energy for the manufacturing chain would drastically decrease the ecological footprint of the automotive industry.

Recently, Kozakevich et al. [10,14] investigated the interplay between the cooling rates, microstructures, and mechanical properties of an Al-Ce-Ni-Mn alloy designed for elevated-temperature applications. The alloy was cast in a wedge mold configuration to investigate the effects of different cooling rates on the material's properties. The alloy exhibited UTS and YS at ambient temperatures ranging from 107 to 131 MPa and 64 to 81 MPa, respectively. At 250 °C, the alloy retained 75 to 83% of its UTS and 73 to 97% of its YS, even after conditioning at 250 °C for 150 h. The variation in tensile strength and tensile retention at 250 °C was highly dependent on the solidification kinetics of the alloy. Scanning electron microscopy (SEM) and energy-dispersive X-ray spectroscopy (EDS) revealed that the volume fraction, size, and morphology of the five phases found, i.e., α-Al, $Al_{23}Ce_4Ni_6$, $Al_{11}Ce_3$, $Al_{10}CeMn_2$, and $Al_{20}CeMn_2$, were highly dependent on the solidification kinetics and induced directional solidification.

To better understand the solidification kinetics of the Al-Ce-Ni-Mn alloy, differential scanning calorimetry (DSC) was performed on four samples from critical areas of the wedge mold in [26] (these results are summarized further in Section 3.2 of this paper). The experiment concluded that further investigation into the solidification kinetics of this alloy is warranted; a better understanding would enable a broader utilization of the alloy for various elevated-temperature applications in the automotive industry.

Therefore, the present work aims to develop a more comprehensive understanding of the alloy's solidification kinetics by performing more in-depth studies on the same samples from [26] and wedge mold casting from [10,14]. The Al-Ce-Ni-Mn alloy's solidification kinetics, microstructure, and directional solidification-induced composition variances are examined using thermodynamic simulation software, SEM, EDS, optical microscopy, DSC thermal analysis, and inductively coupled plasma–optical emission spectroscopy (ICP-OES).

## 2. Materials and Methods

The Al-Ce-Ni-Mn ingots used in this research were provided by Eck Industries Inc., Manitowoc, WI, USA. The alloy was melted down in an electrical resistance furnace under argon coverage and then cast in a steel wedge mold. The samples, their locations, and other relevant information are provided in Figure 1. A comprehensive description of the preliminary results and details of the casting process is presented in [10]. The main casting parameters were that the alloy was poured at 700 °C and the wedge mold was preheated to 250 °C.

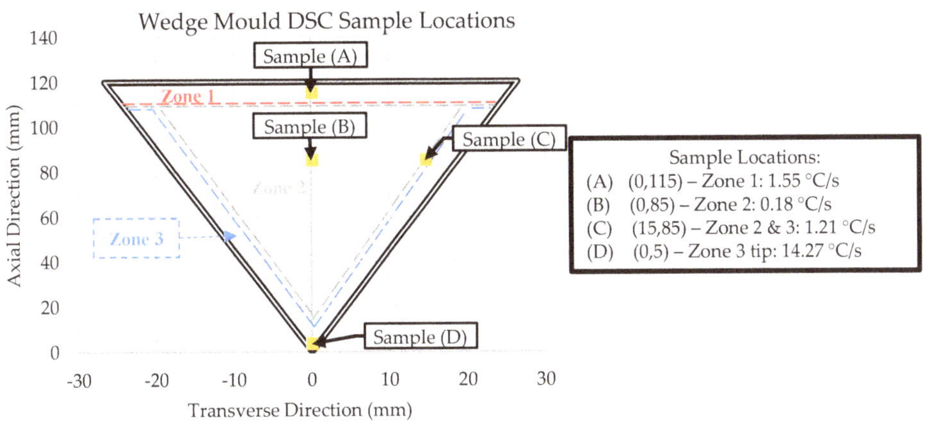


**Figure 1.** Sample details for scanning electron microscopy (SEM), energy dispersive x-ray spectroscopy (EDS), differential scanning calorimetry DSC, and inductively coupled plasma–optical emission spectroscopy (ICP-OES) analysis. NOTE: Sample cooling rate data adapted from [10].

The zones in Figure 1 represent regions with unique microstructures throughout the wedge mold casting. Samples were obtained from critical points within the wedge mold to understand the solidification kinetics of these zones. The critical point in Zone 1 (Sample A) was at the top–center of the wedge mold, where a thermocouple was placed during casting. This sample provides us with the exact cooling rate of the sample to compare with the follow-up DSC and ICP-OES experiments. Similarly, Sample D was taken from the thermocouple placed in Zone 3 for the same reason.

Two samples from Zone 2 were taken. The first (Sample B) was from the center, and the second (Sample C) was from the interface between Zones 2 and 3. Both samples were taken at the same axial position (85 mm), but 15 mm apart in the transverse direction. The center of Zone 2 (Sample B) exhibited the lowest cooling rate (0.18 °C/s). The interface between Zone 2 and 3 (Sample C) provided insight into the transition between the initial skin developed immediately after pouring (high cooling rates) and the center of the wedge mold (lowest cooling rates).

Metallography samples were mounted and prepared according to ASTM E3-11 specifications [27]. Mounting was performed in 2-part fast-curing acrylic. After mounting, the samples were sequentially ground with 400, 600, and 1200 silicon carbide abrasive paper and then polished sequentially with 9, 6, 3, and 1 μm abrasive diamond suspension paste. The final polish was performed with 0.6 μm basic colloidal silica diamond polishing lubricant with a pH of 9.

Once the samples were prepared, optical microscopy was performed to measure the secondary dendrite arm spacing (SDAS). SDAS measurements were taken with a VHX-7000 series digital microscope (KEYENCE, Mississauga, ON, Canada) according to method E outlined in [28], which involved measuring between two secondary dendrites parallel to the primary arm. A minimum of 50 measurements were taken for each sample to ensure measurement and standard deviation accuracy. An example of these measurements is shown in Figure 2.

**Figure 2.** An optical micrograph showing how the secondary dendrite arm spacing (SDAS) measurements were taken.

The SDAS measurements were then used in Equation (1), adapted from [10], which was specific to this material. Utilizing this equation allowed for calculating the cooling rates where thermocouples were absent during casting (i.e., Samples B and C).

$$SDAS = 11.65 * (CR)^{-0.35} \quad (1)$$

A Mira 3 XMU, field emission gun SEM (TESCAN, Brno, Czech Republic) was utilized to collect micrographs of the alloys' microstructures. A Cressington 208 HR sputter coater was used to coat the sample with a 5 μm platinum (Pt) layer to enhance conductivity and reduce the charging effects and heating of the samples. The SEM was operated at an accelerating voltage of 20 kV to capture the micrographs. After the micrographs were collected, Oxford Instruments Aztec data acquisition and processing software (4.3, Oxford Instruments NanoAnalysis, High Wycombe, UK) equipped with an 80 $mm^2$ Oxford EDS detector were utilized to characterize the compositions of each phase. EDS point analysis was performed to identify the atomic percentages of the phases. At least 30 point scans per phase were taken to determine the atomic percentage. The EDS data and phase morphologies were compared with those in the literature to determine each phase's stoichiometry.

A STA 449F3 Jupiter (NETZSCH, Waldkraiburg, Bavaria, Germany) heat flux DSC thermal analysis instrument was employed to understand the Al-Ce-Ni-Mn alloy's phase evolution as a function of temperature. The initial liquidus and solidus temperatures were determined by Scheil ThermoCalc$^{TM}$ simulations to be 733 °C and 632 °C, respectively. Therefore, the temperature range of the DSC experiment was set from 20 to 750 °C. The heating and cooling rate was selected to be 10.00 °C /min (~0.17 °C /s) based on similar studies previously conducted on Al-Ce alloys [18,29]. Each sample was 0.020 +/− 0.003 g and taken from the locations outlined in Figure 1. Samples from the original ingot material were also extracted and tested. A sample size of 0.020 g has been proven to produce the most accurate results for micro-scale DSC thermal analysis [30]. Each sample was heated and cooled three times, and the results were averaged.

The results of the DSC experiment offer insight into the specifics of solidification kinetics, such as the solidus, liquidus, and latent heat of fusion of the eutectic reaction. The solidus and liquidus temperatures were determined by extrapolating the tangent reaction rate near the tip of the characteristic peak and projecting it to intersect the baseline of the DSC data [18,31,32]. The extrapolation of the onset and end temperatures of the characteristic peak was performed according to ASTM E794 [32] standards. The end temperature of the characteristic peak only correlated with the liquidus if one peak was present in the results (i.e., all phases formed near the eutectic temperature). Finally, the latent heat of fusion of the characteristic peak could be calculated by deriving the area under the peak.

ICP-OES (iCAP 6000, Thermo Fisher Scientific, Waltham, MA, USA) was used in this study to confirm the ingot composition and identify the compositional changes that were the result of the induced directional solidification. Samples of 0.1 g each were selected from six random locations on two different ingots. The samples were dissolved in a solution of HCl and $HNO_3$ for digestion while completely retaining the elements, after which the solution was diluted to perform the final elemental analysis. The average composition of the ingots is shown in Table 1.

**Table 1.** Composition of ingots in wt.% based on ICP-OES experimentation.

| Al | Ce | Ni | Mn | Fe |
|---|---|---|---|---|
| Bal. | 12.37 +/−1.72 | 3.26 +/−0.13 | 0.94 +/−0.04 | 0.12 +/−0.03 |

Following this, three samples of 0.1 g from each sample location (A, B, C, and D, identified in Figure 1) were extracted and tested via ICP-OES to determine the difference in composition due to directional solidification in the wedge mold. The average of the three tests was taken as the sample composition for its specific location.

## 3. Results and Discussion

### *3.1. ThermoCalc$^{TM}$ Scheil Solidification*

A Scheil solidification simulation was conducted using the ThermoCalc$^{TM}$ software (2022b, Thermo-Calc Software, Solna, Sweden) based on the composition obtained from the ICP-OES analysis of the ingot material. The results are shown in Figure 3.

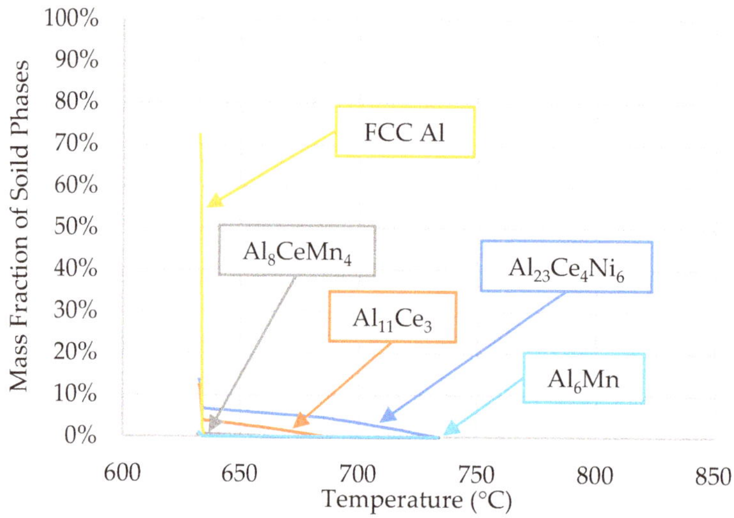


**Figure 3.** Scheil simulation of alloys' solidification for the compositions presented in Table 1.

The phases predicted by ThermoCalc™ were α-Al (FFC Al), $Al_{23}Ce_4Ni_6$, $Al_{11}Ce_3$, $Al_8CeMn_4$, and $Al_6Mn$. The presence of these phases, except for the $Al_6Mn$ phase, was confirmed via SEM/EDS in [10]. Limited research is available on the $Al_{23}Ce_4Ni_6$ phase regarding its solidification kinetics. In the early 2000s, Gout et al. [33] concluded that the $Al_{23}Ce_4Ni_6$ phase crystallizes in the monoclinic space group C2/m, the cell parameters for which are $a$ = 16.042 (8), $b$ = 4.140 (4), c = 18.380 (8) Å, and $\beta$ = 113.24 (5)°. Typically, this ternary phase is accompanied by $Al_{11}Ce_3$ and $Al_3Ni$ in ternary Al-Ce-Ni alloys; however, the results from [10] and ThermoCalc™ were in agreement that $Al_3Ni$ did not form in the Al-Ce-Ni-Mn alloy.

In the preliminary analysis [10], when investigating the interplay between the cooling rate, microstructure, and tensile properties of the current Al-Ce-Ni-Mn alloy, it was found that $Al_8CeMn_4$ was absent from the casting. Instead, it was replaced by $Al_{10}CeMn_2$ and $Al_{20}CeMn_2$. Over the past few years, research has been conducted on Al-Ce-Mn alloys, which has focused on the Al-rich portion of this system [34–36]. Most significant to the solidification kinetics of the Al-Ce-Ni-Mn alloy described in the current study is the revised portion of the Al-rich corner of the Al-Ce-Mn ternary diagram performed by Yang et al. in [34]. This revised portion of the ternary diagram shows that the $Al_{10}CeMn_2$ and $Al_{20}CeMn_2$ phases dominated earlier during solidification, with the latter phase having a slightly higher thermodynamic driving force. From this revised ternary diagram, it is expected, that for the composition of the alloy under investigation (12.37% Ce, 0.94% Mn), the $Al_{10}CeMn_2$ phase would form earlier during solidification, after which, at ~700 °C, the ternary Al-Ce-Mn phase would transition to $Al_{20}CeMn_2$ and $Al_{11}Ce_3$. This phenomenon is shown below in Figure 4.

Aside from Scheil simulations, ThermoCalc was also used to calculate the solubility of the elements in each phase. More specifically, the amounts of solid-solution Mn in $Al_{23}Ce_4Ni_6$, Ni and Mn in $Al_{11}Ce_3$, Ni in $Al_8CeMn_4$, and Mn in the α-Al matrix were determined throughout the solidification range (734 to 632 °C). Negligible solubility was found in all cases, except for Mn in the α-Al matrix, where the maximum solid solution at 632 °C was 0.4 at.%, lower than the 0.6 at.% published in the binary Al-Mn literature [37,38]. However, Mondolfo's investigation into Al-Ce-Mn alloys [39] revealed that Ce reduces Mn's solid solubility in Al. The SEM/EDS data discussed in the next section show how much Mn is present in the Al matrix and how its solubility is affected by the solidification rate of the alloy.

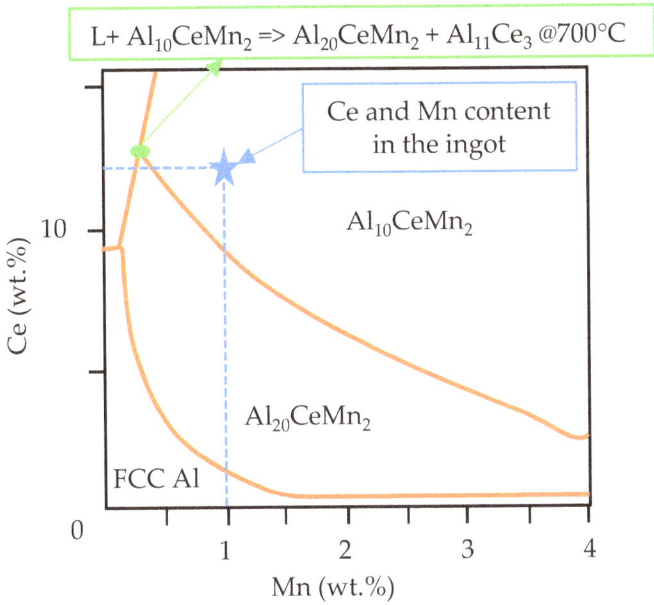


**Figure 4.** The modified ternary Al-Ce-Mn liquidus projection and the position of the Al-Ce-Ni-Mn alloy investigated in this stud, data adapted from [34].

### 3.2. SEM/EDS Solidification Trends

Previously in [10], the resulting wedge casting revealed three distinct zones with varying microstructures. Figure 1 shows the sample locations and how the respective cooling rates were determined, while Figure 5 below shows the microstructures at those sample locations. The phases in each zone were relatively similar and contained variations of α-Al, eutectic Al-$Al_{23}Ce_4Ni_6$, eutectic Al-$Al_{11}Ce_3$, primary $Al_{23}Ce_4Ni_6$, primary $Al_{11}Ce_3$, and $Al_{20}CeMn_2$. The $Al_{10}CeMn_2$ phase was only present in samples C and D near the wall of the casting. All Mn in samples A and B was either in solid solution in the Al matrix or $Al_{20}CeMn_2$ phase.

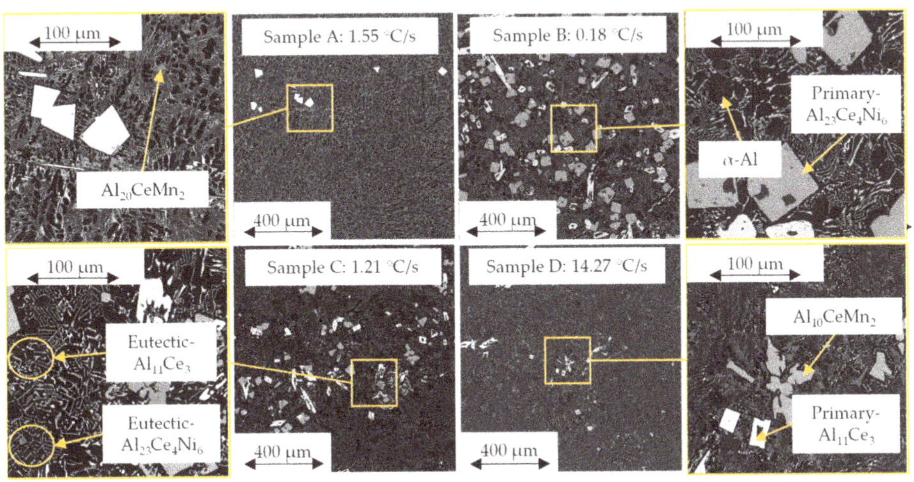


**Figure 5.** SEM micrographs of each sample's general microstructure.

Point scans were performed for each phase to determine the atomic composition. After this, the atomic composition was compared with that in the published literature on similar phases to determine the correct stoichiometry of a specific phase. The atomic compositions of each phase are summarized in Table 2.

**Table 2.** The atomic percentages of phases identified in the Al-Ce-Ni-Mn alloy by EDS.

| Phase Identification | Element in Phase (at.%) | | | | |
|---|---|---|---|---|---|
| | Al | Ce | Ni | Mn | Fe |
| Matrix (α-Al) | 99.77 +/−0.21 | - | - | 0.23 +/−0.21 | - |
| $Al_{11}Ce_3$—Primary | 79.61 +/−1.37 | 20.36 +/−1.36 | - | - | - |
| $Al_{11}Ce_3$—Eutectic | 93.43 +/−1.42 | 6.41 +/−1.53 | 0.16 +/−0.37 | - | - |
| $Al_{23}Ce_4Ni_6$—Primary | 73.37 +/−0.57 | 8.19 +/−0.11 | 15.77 +/−0.34 | 2.56 +/−0.06 | 0.01 +/−0.05 |
| $Al_{23}Ce_4Ni_6$—Eutectic | 89.09 +/−3.14 | 3.25 +/−0.99 | 6.57 +/−1.93 | 1.10 +/−0.24 | - |
| $Al_{10}CeMn_2$ | 79.02 +/−0.83 | 7.47 +/−0.33 | 3.33 +/−0.20 | 10.17 +/−0.45 | - |
| $Al_{20}CeMn_2$ | 87.99 +/−1.00 | 4.20 +/−0.27 | 0.32 +/−0.35 | 7.50 +/−0.77 | - |

The Al matrix contained an average of 0.23 atomic percentage (at.%) Mn in solid solution. Higher cooling rates near the wall (e.g., Sample D) resulted in up to 0.44 at.% Mn in the matrix. As previously stated, these values are lower than those in the published literature [37,38] on Al-Mn alloys (~0.62 at.% Mn into α-Al). However, as also mentioned above, Mondolfo's investigation into Al-Ce-Mn alloys [39] revealed that Ce lowered the solubility of Mn in the matrix, which explained the lower values of Mn in α-Al presented in Table 2.

$Al_{11}Ce_3$ was present in two morphologies: blocky primary and lamellae eutectic phases. These morphologies are consistent with those in the literature data on the binary Al-Ce alloys [17,19,20]. The primary intermetallic phase resided in samples B, C, and D, but not at the top of the casting (Sample A). The stoichiometry of the primary phase was consistent with the literature [40] and contained no Ni or Mn in a solid solution. The lamellae eutectic $Al_{11}Ce_3$ contained a trace amount of Ni (0.16 at.%) in a solid solution. Research by Tang et al. in [41,42] revealed that between 0.6 and 1.7 at.% Ni can be dissolved into the $Al_{11}Ce_3$ phase. However, the alloys in these studies had significantly higher Ce and Ni contents, which justifies why they experienced an elevated amount of Ni in the solid solution of the $Al_{11}Ce_3$ phase.

The solubility of Ni in the eutectic phase and not the primary phase can be explained by the precipitation kinetics of the $Al_{23}Ce_4Ni_6$ phases. In the early stages of solidification (~732 °C), the primary $Al_{23}Ce_4Ni_6$ phase (also known as the τ8 phase [33,41–44]) began to form, drawing Al and Ni out of the liquid at a higher rate than Ce. Eventually, the liquid became Ce-enriched, and the $Al_{11}Ce_3$ primary phase began to precipitate out of the liquid around 686 °C. As the primary $Al_{23}Ce_4Ni_6$ phase had a high growth rate, no excess Ni could be dissolved into the primary $Al_{11}Ce_3$ phase. However, once the alloy reached the eutectic temperature, the precipitation of the lamellae Al-$Al_{23}Ce_4Ni_6$ and Al-$Al_{11}Ce_3$ phases occurred simultaneously. These phases were thin and fine, and because Ni had negligible solubility in the Al matrix, the remaining Ni entered both eutectic lamellae phases, resulting in ~0.16 at.% Ni in the Al-$Al_{11}Ce_3$ eutectic phase. However, due to the nano-scale size of the $Al_{11}Ce_3$ eutectic phase, the electron beam during EDS analysis may have been penetrating the surrounding Al-matrix, thereby diluting the measurements. Therefore, the expected Ni composition in the eutectic Al-$Al_{11}Ce_3$ could be calculated based on the ratio of Ce to Ni. As the Ce:Ni ratio was ~40:1, we could scale the 6.41 at.% Ce up to ~20 at.% based on our EDS data, which is consistent with the literature. Doing so scaled the Ni at.% in the eutectic $Al_{11}Ce_3$ to ~0.509 at.%, which was closer to the values published in the literature [41,42].

Similarly, the EDS results from the $Al_{23}Ce_4Ni_6$ eutectic phase also presented evidence of dilution due to the surrounding Al-matrix. When scaling the Ni:Mn ratio of the eutectic phase up to the expected composition of 15.77 at.% Ni, it is expected that ~2.6 at.% Mn would be found in the solid solution of the eutectic $Al_{23}Ce_4Ni_6$ lamellae, which was approximately the Mn content found in the primary phase. Aside from the Mn content found in the $Al_{23}Ce_4Ni_6$ phases, the amounts of Ce and Ni shown in Table 2 are consistent with the published data in [33,41–44]. The eutectic and primary phases were larger in sample B than those in the other samples. This is consistent with the results presented in [14]. As shown in Table 2, both morphologies of this phase had significant Mn solubility (~2.56 at.%), a phenomenon that is not currently predictable by the ThermoCalc™ software.

The same conclusion applies to both ternary Al-Ce-Mn phases in the wedge mold casting. It was found that the $Al_{10}CeMn_2$ phase had an average of 3.3 at.% Ni in solid solution, while the $Al_{20}CeMn_2$ phase contained much less (i.e., 0.32 at.%) Ni in solid solution. The $Al_{10}CeMn_2$ and $Al_{20}CeMn_2$ phases were consistent in size and morphology with those found and characterized in [34,35].

### 3.3. DSC Thermal Analysis

The SEM micrographs and EDS compositional analysis show that the phase morphologies and volume fractions within each zone of the wedge mold casting differed significantly. These factors depended on the changes in the progression of phase evolution that resulted from varying cooling rates. DSC thermal analysis was necessary to understand the phase precipitation characteristics and to gain further insight into the solidification kinetics of this alloy. Initial DSC thermal analysis was performed in [26] to show how the solidification kinetics affected the alloy's microstructure. The preliminary DSC results are shown in Figure 6 below. The indicated sample locations correspond to those in the experimental procedure section above. The solidus temperature, latent heat of fusion of the eutectic phase, and liquidus temperature were extracted from the DSC results.

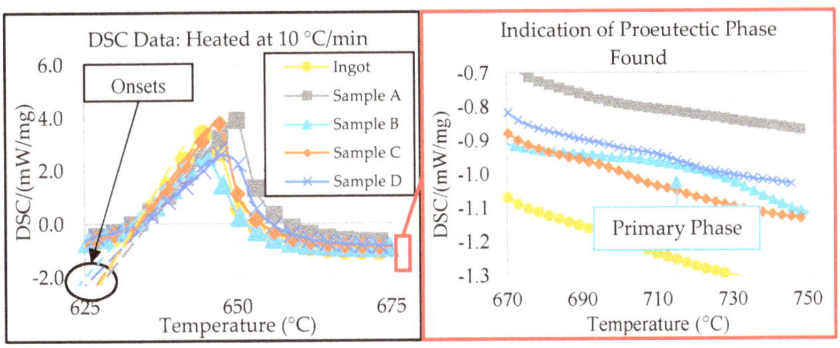


**Figure 6.** DSC data of all samples, modified and adapted from [26].

Figure 6 shows that all samples provided a similar solidus temperature (onset) to the 632 °C predicted by the Scheil simulation of the ingot material in ThermoCalc™, except for that of Sample B, which was 5 °C lower [26]. The discrepancy in eutectic temperatures between Sample B and the ingot suggested a compositional variance in the center of the wedge mold (Zone 2). Additionally, the exploded view in Figure 6, on the right, shows evidence of primary formation in samples B, C, and D at different temperatures, suggesting compositional variations between the DSC samples. These results are consistent with the SEM micrographs.

The compositional difference between the samples, indicated by the initial DSC results, suggested that further investigation was needed before determining if the end temperature of the eutectic reaction could be considered as the liquidus temperature for Samples A, C, and D. Additional information is also needed from sample B to ensure that no other primary peak forms above 800 °C. For this reason, the determined liquidus temperature of each sample is discussed in Section 3.4 below.

Aside from the liquidus and solidus temperatures, the DSC data show the significant variance in the eutectic reaction latent heat of fusion between samples. The latent heat of fusion from DSC thermal analysis was calculated based on the energy absorbed (exothermic is identified in Figure 6) during the melting of the eutectic phase, which was then normalized by the sample's total mass. Therefore, if the samples had significantly different latent heat of fusion values, they must also have had different volume fractions of eutectic phases. For example, sample B contained the least characteristic latent heat of fusion. In Figure 5, sample B also had the highest volume fraction of primary phases, indicating that sample B had a different composition to the original ingot. This compositional difference resulted from the cooling gradient induced by the wedge mold configuration. This significant cooling gradient varied the alloy's solidification kinetics, resulting in differences in the compositions within the critical zones of the casting. The significance of how the composition varied was characterized via ICP-OES analysis.

### 3.4. ICP-OES Data

ICP-OES analysis was performed to characterize the composition gradient that directly resulted from the solidification kinetics of the Al-Ce-Ni-Mn alloy and the directional solidification induced by the wedge mold casting configuration. The results of the ICP-OES experiments are shown in Figure 6. In Figure 6, the downward-pointing arrows highlighted in red indicate the elemental percentage of change compared with the original ingot's composition shown in Table 1. The green upward-pointing arrows indicate increases in elemental composition based on the original ingot values.

Figure 7 shows that the ICP-OES results of samples A and C exhibited significant decreases in Ce, Ni, and Mn compared with the original ingot composition. In turn, the Ce, Ni, and Mn concentrations increased in Samples B and D. The exact compositions are shown below in Table 3.

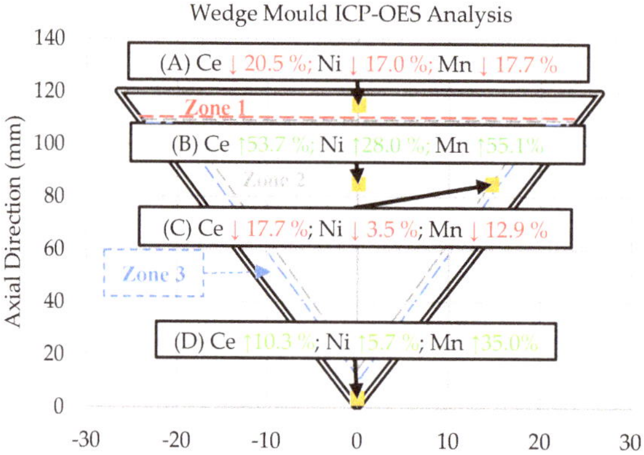


**Figure 7.** Summary of the wt.% changes of each sample compared with the ingot composition as determined by the ICP-OES analysis of locations in the wedge mold. NOTE: The up and down arrows indicate percent increases or decreases, respectively, from the original ingot composition in Table 1.

**Table 3.** ICP-OES data for the Ingot materials and samples A, B, C, and D (all values in wt.%).

| | Al | Ce | Ni | Mn | Fe |
|---|---|---|---|---|---|
| Ingot (Reference) | Bal. | 12.374 +/− 1.722 | 3.263 +/− 0.129 | 0.937 +/− 0.041 | 0.119 +/− 0.032 |
| (A) 1.55 °C/s | Bal. | 9.842 +/− 0.185 (−20.5%) | 2.707 +/− 0.027 (−17.0%) | 0.771 +/− 0.016 (−17.7%) | 0.0917 +/− 0.002 (−22.8%) |
| (B) 0.18 °C/s | Bal. | 19.024 +/− 0.712 (+53.7%) | 4.175 +/− 0.192 (+28.0%) | 1.452 +/− 0.017 (+55.1%) | 0.157 +/− 0.002 (+32.0%) |
| (C) 1.21 °C/s | Bal. | 10.183 +/− 0.128 (−17.7%) | 3.150 +/− 0.032 (−3.5%) | 0.816 +/− 0.012 (−12.9%) | 0.087 +/− 0.001 (−26.6%) |
| (D) 14.27 °C/s | Bal. | 13.652 +/− 0.104 (+10.3%) | 3.450 +/− 0.028 (+5.7%) | 1.264 +/− 0.015 (+35.0%) | 0.157 +/− 0.001 (+32.2%) |

The ICP-OES-determined compositions of each sample were used in Scheil simulations to determine the solidification kinetics of each critical zone of the wedge mold. The results of these Scheil simulations are shown in Figure 8.

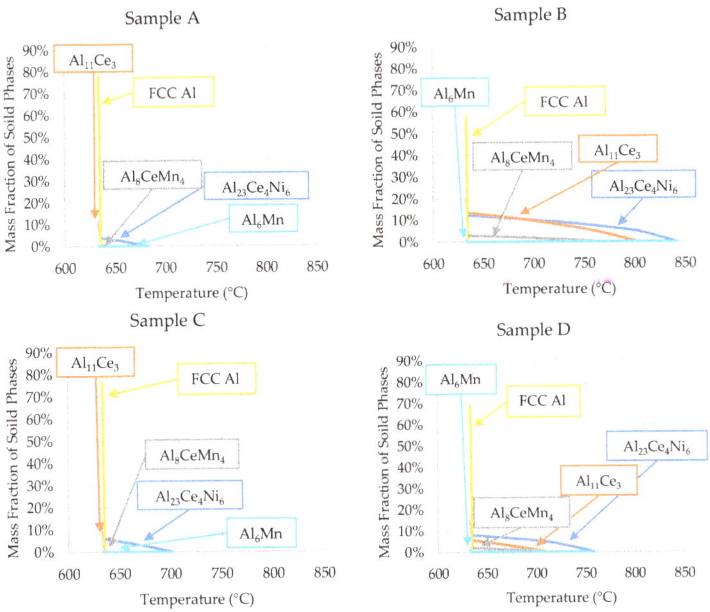


**Figure 8.** Scheil simulations of the alloys' solidification for the compositions determined by the ICP-OES analyses of each sample position (presented in Table 3).

The phase evolutions shown in Figure 8 are consistent with the SEM micrographs and DSC data. The only discrepancy is that small amounts of $Al_6Mn$ precipitated close to the end of Scheil solidification in all the samples. However, SEM/EDS showed no presence of $Al_6Mn$ in the casting. As mentioned, ThermoCalc$^{TM}$ did not account for the precipitation of $Al_{10}CeMn_2$ and $Al_{20}CeMn_2$. Instead, the software predicted the $Al_8CeMn_4$ phase to precipitate, which was disproved by the EDS data in Table 2.

The Scheil simulation for each sample revealed that Sample A contained proeutectic $Al_{23}Ce_4Ni_6$ starting to precipitate at 682 °C. The software predicted 3.8 wt.% (<1 vol.%) of the primary $A_{23}Ce_4Ni_6$ phase, which was consistent with the SEM micrographs. The relatively low amount of the primary phase, coupled with the precipitation temperature being close to the eutectic temperature, explains why the phase did not register in the DSC results. It is likely that, during DSC experimentation, the proeutectic formation of the $Al_{23}Ce_4Ni_6$ phase was encapsulated in the characteristic peak. Therefore, the end temperature of the characteristic peak (655 °C) from the DSC results for sample A was likely the liquidus temperature of the sample. This value was lower than the Scheil-predicted value of 682 °C. The disagreement between liquidus temperatures was likely because the mass fraction of the primary phases was not significant enough for the DSC to detect the resulting reaction energy. The DSC and Scheil simulation agreed that the solidus temperature of sample A was 632 °C.

The comparison between the DSC and ThermoCalc$^{TM}$ results is summarized in Table 4 for all samples. Table 4 shows that all samples agreed with the Scheil simulation and DSC results of the solidus temperature, except for sample B. The DSC results for Sample B showed that the solidus temperature was 627 °C, while the Scheil simulation predicted the solidus temperature to be 632 °C, a 5 °C difference. The other samples were between 1 and 3 °C off the simulated value.

**Table 4.** Liquidus and solidus temperatures of ICP-OES sample compositions based on Scheil simulations compared with DSC.

| | **Liquidus (°C)** | | **Solidus (°C)** | |
|---|---|---|---|---|
| | **DSC** | **ThermoCalc$^{TM}$** | **DSC** | **ThermoCalc$^{TM}$** |
| Ingot | 662 | 734 | 632 | 632 |
| Sample A | 655 | 682 | 633 | 632 |
| Sample B | - | 841 | 627 | 632 |
| Sample C | 700 | 701 | 632 | 632 |
| Sample D | - | 758 | 629 | 632 |

Note: "-" indicates that the value could not be interpreted from the DSC results because it was outside the original solidification range predicted by the ingot composition and Scheil simulation.

The solidification range of sample B's ICP-OES composition increased significantly due to the hypereutectic Ce and Ni contents. The precipitation of the primary $Al_{23}Ce_4Ni_6$ phase began at 840 °C, compared with 734 °C in the ingot material, as predicted by the Scheil simulation. As the DSC experiment on this sample only went up to 800 °C, the liquidus temperature of the alloy could not be determined by the DSC data. However, the Scheil simulation of sample B's composition in Figure 8 offers insight into what exactly is happening during solidification within the wedge mold. The large solidification range and low cooling rate (0.18 °C/s) explain why the SEM micrographs of Sample B showed larger primary phases than the other samples. As Sample B was from Zone 2, it could be observed that, when Zone 3 solidified (sample D) first at a relatively high cooling rate (14.27 °C/s), elevated amounts of Ce and Ni resided in the liquid. Excess Ni and Ce resided in the liquid because the high cooling rate of Zone 3 did not allow enough time for the primary $Al_{23}Ce_4Ni_6$ and $Al_{11}Ce_3$ phases to precipitate in significant amounts. Zone 2 began to solidify much slower because the steel mold was heat-soaked, resulting in significantly reduced heat flux. The slower solidification rate and elevated amounts of Ce and Ni allowed for the formation of large primary $Al_{23}Ce_4Ni_6$ and $Al_{11}Ce_3$ intermetallics.

Sample D's Ni and Mn contents were very similar to those of sample B; however, the microstructure was vastly different. Figure 8 and Table 4 show that the solidification range for sample D compared with sample B was ~83 °C smaller. This smaller solidification range and the significantly higher cooling rate resulted in the precipitation of the $Al_{10}CeMn_2$ phase in sample D, but not in sample B. As previously stated, the revised Al-Ce-Mn ternary diagram from Yang et al. [34] shows that the ternary $Al_{10}CeMn_2$ phase transitioned to $Al_{11}Ce_3$ and $Al_{20}CeMn_2$ at 700 °C. The high cooling rates experienced in sample D did not allow enough time for this reaction to occur, trapping the $Al_{10}CeMn_2$ phase in the solid. This phenomenon was also evident in sample C at a cooling rate of 1.21 °C/s. Samples A and B had no $Al_{10}CeMn_2$ in their microstructure, which indicates that, to remove this phase, an Mn content of 0.771 wt.% (or less) or a cooling rate lower than 1.21 °C/s is required.

Sample C showed good agreement in the solidus and liquidus temperatures between the Scheil simulation and DSC results, as shown in Table 4. As sample C had a composition closely resembling the ingot composition, the formation of the primary phases was presented in the DSC results. Therefore, the liquidus temperature (700 °C) could be extracted from the DSC results.

## 4. Conclusions

This study investigated the solidification kinetics of an Al-Ce alloy with additions of Ni and Mn when cast in a wedge mold configuration. The wedge mold configuration was determined to induce significant directional solidification, leading to large microstructure variations throughout the final cast product. When analyzing the three critical zones of the wedge mold casting via SEM/EDS, DSC Thermal Analysis, and ICP-OES, the following conclusions were drawn:

1. The cooling rate and directional solidification drastically affected the formation of the ternary Al-Ce-Mn phases. Cooling rates above 1.21 °C/s did not allow the $Al_{10}CeMn_2$ phase to transition into $Al_{20}CeMn_2$ + $Al_{11}Ce_3$ at 700 °C. Therefore, the Al-Ce-Ni-Mn alloy can have its mechanical properties tailored based on which Al-Ce-Mn ternary phase benefits specific applications.
2. Induced directional solidification also significantly impacted the consistency of composition throughout the final cast product. The solidification sequence must be considered when casting these novel alloys, as rapidly solidified areas will lead to hypereutectic compositions in the slower-cooled regions of the casting, as is evident in the ICP-OES results in this study. A hypereutectic composition and large solidification range lead to significant primary blocky $Al_{23}Ce_4Ni_6$ and $Al_{11}Ce_3$ phases, which may be desirable or undesirable, depending on the application.
3. The $Al_{23}Ce_4Ni_6$ phase had a notably higher (~2.6 at.%) Mn solubility than that predicted by the ThermoCalc$^{TM}$ software. The same was true for the solubility of Ni in $Al_{10}CeMn_2$ (3.33 at.%) and $Al_{20}CeMn_2$ (0.32 at.%).

The comprehensive correlations between the solidification cooling rate, directional solidification-induced compositional variances, and the resulting microstructure/phase evolution provide critical insight into how this novel Al alloy can be tailored for different applications in the automotive industry.

**Author Contributions:** Conceptualization, J.R.K.; methodology, J.R.K. and J.S.; software, J.R.K.; validation, J.R.K. and J.S.; formal analysis, J.R.K.; investigation, J.R.K.; resources, D.S. and D.W.; data curation, J.R.K.; writing—original draft preparation, J.R.K.; writing—review and editing, J.S., D.S. and D.W.; visualization, J.R.K. and J.S.; supervision, D.S. and D.W.; project administration, D.S. and D.W.; funding acquisition, D.S. and D.W. All authors have read and agreed to the published version of the manuscript.

**Funding:** This work was supported by Eck Industries and the Mitacs International Award (Application Ref. IT23293).

**Data Availability Statement:** The raw and processed data required to produce these findings are available for download from Mendeley Data "Solidification kinetics of an Al-Ce alloy with additions of Ni and Mn", https://data.mendeley.com/datasets/nmmvwn9yf4, Accessed April 2023.

**Acknowledgments:** The authors would like to acknowledge Mark Button and Sudip Shrestha of the UBCO FiLTER laboratory for their contributions to the ICP-OES and SEM data used in this article.

**Conflicts of Interest:** David Weiss is an employee of Eck Industries Inc. and is one of the inventors of the alloy system.

## References

1. Castro-Alvarez, A.; Gil, Y.; Llanos, L.; Aravena, D. High performance single-molecule magnets, Orbach or Raman relaxation suppression. *Inorg. Chem. Front.* **2020**, *7*, 2478–2486. [CrossRef]
2. Goll, D.; Kronmüller, H. High-performance permanent magnets. *Naturwissenschaften* **2000**, *87*, 423–438. [CrossRef] [PubMed]

3. Hirosawa, S.; Nishino, M.; Miyashita, S. Perspectives for high-performance permanent magnets: Applications, coercivity, and new materials. *Adv. Nat. Sci. Nanosci. Nanotechnol.* **2017**, *8*, 013002. [CrossRef]
4. Survey, U.S.G. Mineral Commodity Summaries. Available online: https://www.usgs.gov/centers/nmic/mineral-commodity-summaries (accessed on 15 January 2023).
5. Sims, Z.C.; Kesler, M.S.; Henderson, H.B.; Castillo, E.; Fishman, T.; Weiss, D.; Singleton, P.; Eggert, R.; McCall, S.K.; Rios, O. How cerium and lanthanum as coproducts promote stable rare earth production and new alloys. *J. Sustain. Metall.* **2022**, *8*, 1225–1234. [CrossRef]
6. Belov, N.A.; Naumova, E.A.; Eskin, D.G. Casting alloys of the Al-Ce-Ni system: Microstructural approach to alloy design. *Mater. Sci. Eng. A* **1999**, *271*, 134–142. [CrossRef]
7. Stroh, J.; Sediako, D.; Weiss, D. Development of cerium-reinforced specialty aluminum alloy with application of X-ray and neutron diffraction. *Int. J. Met.* **2021**, *15*, 29–39. [CrossRef]
8. Aghaie, E.; Stroh, J.; Sediako, D.; Rashidi, A.; Milani, A.S. Improving the mechanical properties of the B319 aluminum alloy by addition of cerium. *Mater. Sci. Eng. A* **2020**, *793*, 139899. [CrossRef]
9. Kozakevich, J.R.; Stroh, J.; Sediako, D.; Weiss, D.; Loukus, A.; Vogel, S.C. Elevated temperature creep and tensile performance of extruded Mg-10Ce alloy. *J. Mater. Eng. Perform.* **2022**, *32*, 2758–2765. [CrossRef]
10. Kozakevich, J.R.; Stroh, J.; Mallouhi, V.; Sediako, D.; Weiss, D. Interplay between cooling rate, microstructure, and mechanical properties of an Al-Ce-Ni-Mn alloy. In *Light Metals*; Springer: Berlin/Heidelberg, Germany, 2022; pp. 83–89. [CrossRef]
11. Fodran, E.J. Microstructural Evolution and Thermal Stability of Al-Ce-Ni ternary Eutectic. Ph.D. Thesis, The University of Florida, Gainesville, FL, USA, August 2002.
12. Weiss, D. Improved high-temperature aluminum alloys containing cerium. *J. Mater. Eng. Perform.* **2019**, *28*, 1903–1908. [CrossRef]
13. Czerwinski, F. Thermal stability of aluminum–cerium binary alloys containing the Al–Al11Ce3 eutectic. *Mater. Sci. Eng. A* **2021**, *809*, 140973. [CrossRef]
14. Kozakevich, J.R. Development and Characterization of A Novel Near-Eutectic Al-Ce Alloy with Additions of Ni and Mn for Elevated Temperature Applications in the Automotive Industry. Master's Thesis, The University of British Columbia, Kelowna, BC, Canada, December 2022.
15. Stroh, J. Development of Orecipitation-Strengthened Aluminum Alloys and Manufacturing Processes for Next Generation Automotive Powertrains. Ph.D. Thesis, The University of British Columbia, Kelwona, BC, Canada, April 2021.
16. Sims, Z.C.; Rios, O.R.; Weiss, D.; Turchi, P.E.; Perron, A.; Lee, J.R.; Li, T.T.; Hammons, J.A.; Bagge-Hansen, M.; Willey, T.M.; et al. High performance aluminum-cerium alloys for high-temperature applications. *Mater. Horiz.* **2017**, *4*, 1070–1078. [CrossRef]
17. Stroh, J.; Sediako, D.; Weiss, D.; Peterson, V.K. In situ neutron diffraction solidification analyses of rare earth reinforced hypoeutectic and hypereutectic aluminum-silicon alloys. In *Light Metals 2020*; Springer: Berlin/Heidelberg, Germany, 2020; pp. 174–178. [CrossRef]
18. Aniolek, M.; Smith, T.; Czerwinski, F. Combining differential scanning calorimetry and cooling- heating curve thermal analysis to study the melting and solidification behavior of al-ce binary alloys. *Metals* **2021**, *11*, 372. [CrossRef]
19. Czerwinski, F.; Amirkhiz, B.S. On the Al-Al11Ce3 eutectic transformation in aluminum-cerium binary alloys. *Materials* **2020**, *13*, 4549. [CrossRef] [PubMed]
20. Czerwinski, F. A search for the eutectic system of high-temperature cast aluminum alloys. *Mater. Sci. Technol.* **2021**, *37*, 683–692. [CrossRef]
21. Sims, Z.C.; Weiss, D.; McCall, S.K.; McGuire, M.A.; Ott, R.T.; Geer, T.; Rios, O.; Turchi, P.A.E. Cerium-based, intermetallic-strengthened aluminum casting alloy: High-volume co-product development. *J. Miner. Met. Mater. Soc.* **2016**, *68*, 1940–1947. [CrossRef]
22. Czerwinski, F. Thermal stability of aluminum alloys. *Materials* **2020**, *13*, 3441. [CrossRef]
23. Weiss, D.; Rios, O.; Sims, Z.C.; Mccall, S.; Ott, R. Casting characteristics of high cerium content aluminum alloys. In *Light Metals*; Springer: Berlin/Heidelberg, Germany, 2017; pp. 205–211.
24. Nguyen, R.T.; Imholte, D.D.; Rios, O.R.; Weiss, D.; Sims, Z.; Stromme, E.; McCall, S.K. Anticipating impacts of introducing aluminum-cerium alloys into the United States automotive market. *Resour. Conserv. Recycl.* **2019**, *144*, 340–349. [CrossRef]
25. Salonitis, K.; Jolly, M.; Pagone, E.; Papanikolaou, M. Life-cycle and energy assessment of automotive component manufacturing: The dilemma between aluminum and cast iron. *Energies* **2019**, *12*, 2557. [CrossRef]
26. Kozakevich, J.R.; Stroh, J.; Sediako, D.; Weiss, D. DSC thermal analysis of a near eutectic Al-Ce alloy with additions of Ni and Mn. *COM Light Met.* **2023**, *in press*.
27. *ASTM(2017)*; Standard Guide for Preparation of Metallographic Specimens. ASTM International: West Conshohocken, PA, USA, 2017. [CrossRef]
28. Vandersluis, E.; Ravindran, R. Comparison of measurement methods for secondary dendrite arm spacing. *Metallogr. Microstruct. Anal.* **2017**, *6*, 89–94. [CrossRef]
29. Zhang, H.; Zhang, C.; Zhang, L.; Zhai, F.; Tian, J. Effect of Cu on the microstructure and the property of Al-Ce-Ni-Cu amorphous nano-composite materials. *Adv. Mater. Res.* **2013**, *624*, 248–251. [CrossRef]
30. Haghdadi, H.; Phillion, A.B.; Maijer, D.M. Microstructure characterization and thermal analysis of aluminum alloy B206 during solidification. *Metall. Mater. Trans. A Phys. Metall. Mater. Sci.* **2015**, *46*, 2073–2081. [CrossRef]

31. Boettinger, W.J.; Kattner, U.R.; Moon, K.W.; Perepezko, J.H. DTA and heat-flux DSC measurements of alloy melting and freezing. In *Methods for Phase Diagram Determination*; Zhao, J.C., Ed.; Elsevier Ltd.: Amsterdam, The Netherlands, 2007; pp. 151–221. [CrossRef]
32. *ASTM (2015) E794-06*; Standard Test Method for Melting and Crystallization Temperatures by Thermal. ASTM International: West Conshohocken, PA, USA, 2015. [CrossRef]
33. Gout, D.; Benbow, E.; Gourdon, O.; Miller, G.J. Crystallographic, electronic and magnetic studies of Ce4Ni 6Al23: A new ternary intermetallic compound in the cerium-nickel-aluminum phase diagram. *J. Solid State Chem.* **2003**, *174*, 471–481. [CrossRef]
34. Yang, Y.; Bahl, S.; Sisco, K.; Lance, M.; Shin, D.; Shyam, A.; Plotkowski, A.; Dehoff, R.R. Primary solidification of ternary compounds in Al-rich Al–Ce–Mn alloys. *J. Alloys Compd.* **2020**, *844*, 156048. [CrossRef]
35. Coury, F.G.; Kiminami, C.S.; Botta, W.J.; Bolfarini, C.; Kaufman, M.J. Design and production of Al-Mn-Ce alloys with tailored properties. *Mater. Des.* **2016**, *110*, 436–448. [CrossRef]
36. Gordillo, M.A.; Cernatescu, I.; Aindow, T.T.; Watson, T.J.; Aindow, M. Phase stability in a powder-processed Al-Mn-Ce alloy. *J. Mater. Sci.* **2014**, *49*, 3742–3754. [CrossRef]
37. Liu, X.J.; Ohnuma, J.; Kainuma, R.; Ishida, K. Thermodynamic assessment of the Aluminum-Manganese (Al-Mn) binary phase diagram. *J. Phase Equilibria* **1999**, *20*, 45–56. [CrossRef]
38. Mondolfo, L.F. Al–Mn Aluminum–Manganese system. In *Aluminum Alloys: Structures and Properties*; Butterworth-Heinemann: Oxford, UK, 1976; pp. 324–329. [CrossRef]
39. Mondolfo, L.F. Al–Ce–Mn Aluminum–Cerium–Manganese system. In *Aluminum Alloys: Structures and Properties*; Butterworth-Heinemann: Oxford, UK, 1976; pp. 470–471. [CrossRef]
40. Okamoto, H. Al-Ce (Aluminum-Cerium). *J. Phase Equilibria Diffus.* **2011**, *32*, 392–393. [CrossRef]
41. Tang, C.; Du, Y.; Zhou, H. The phase equilibria of the Al-Ce-Ni system at 500 °C. *J. Alloys Compd.* **2009**, *470*, 222–227. [CrossRef]
42. Tang, C.; Du, Y.; Xu, H.H.; Xiong, W.; Zhang, L.J.; Zheng, F.; Zhou, H.Y. Experimental investigation of the Al-Ce-Ni system at 800 °C. *Intermetallics* **2008**, *16*, 432–439. [CrossRef]
43. Raghavan, V. Al-Ce-Ni (Aluminum-Cerium-Nickel). *J. Phase Equilibria Diffus.* **2009**, *30*, 265–267. [CrossRef]
44. Tang, C.; Du, Y.; Wang, J.; Zhou, H.; Zhang, L.; Zheng, F.; Lee, J.; Yao, Q. Correlation between thermodynamics and glass forming ability in the Al-Ce-Ni system. *Intermetallics* **2010**, *18*, 900–906. [CrossRef]

metals MDPI

Article

# Experimental Evaluation of Mechanical and Tribological Properties of Segregated Al-Mg-Si Alloy Filled with Alumina and Silicon Carbide through Different Types of Casting Molds

**Mohammed Y. Abdellah [1,2,*], Bandar M. Fadhl [2], H. M. Abu El-Ainin [3], Mohamed K. Hassan [2,3], Ahmed H. Backar [2,4] and Ahmed F. Mohamed [2,5]**

1 Mechanical Engineering Department, Faculty of Engineering, South Valley University, Qena 83523, Egypt
2 Mechanical Engineering Department, College of Engineering and Islamic Architecture, Umm Al-Qura University, Makkah 21955, Saudi Arabia
3 Production Engineering & Design Department, Faculty of Engineering, Minia University, Minia 61111, Egypt
4 Production Engineering Department, Faculty of Engineering, Alexandria University, Alexandria 21544, Egypt
5 Mechanical Engineering Department, Faculty of Engineering, Sohag University, Sohag 82524, Egypt
* Correspondence: myahmad@uqu.edu.sa or mohamed_abdalla@eng.svu.edu.eg

**Abstract:** A 6061 aluminum alloy has almost 0.8–1.2 wt.% Mg and 0.4–0.8 wt.% Si content. These two components, along with other alloying elements, therefore, were characterized by high mechanical and abrasive strength. The aims of the present work were to understand the effect of different types of cooling rates through different molds materials and to investigate the effect of casting with ceramic additives on segregation of the aluminum alloy itself as a composite material forum. Therefore, a series of mechanical tests were conducted, such as compression test, Vickers hardness, and pin-on-disc wear test. The samples were cast at 650 °C and in electric furnaces for 2 h to ensure that the metal achieved adequate homogeneity and temperature. Then, abrasive macroparticles of $Al_2O_3$ and Sic with a size close to 40–60 µm were used. The particles were poured under constant stirring for 1 min. Then, they were cast in two types of molds: steel and graphite. The cast specimens were obtained as a reference without particles and with 0.5 wt.%, 1 wt.%, 2 wt.%, 3 wt.%, 4 wt.%, and 8 wt.%. The thermal effect and the heat due to conduction and radiation were calculated. The maximum compressive strength was found to increase by ≈21% with SiCp casted in graphite molds, and HV was found to increase by ≈29% with SiC casted in graphite molds. The same was found for wear resistance, which became good with SiC casted in graphite molds, and it was generally found that the cooling rate through the mold weakened the alloy due to the segregation effect. The presence of tough particulate through the aluminum matrix barrier created a number of loads. Additionally, the high specific heat of graphite, which plays a dominant role in the slaw cooling rate of casting, led to grain enlargement, whereas the higher cooling rate of steel led to grain refinement. These concepts are the main rules of heat treatments through the casting process itself, and they save time and effort.

**Keywords:** aluminum alloy; graphite mold; steel mold; casting; heat loss

## 1. Introduction

Metal matrix composites (MMCs) are a strong competitor for other monotonic alloys [1]. They have a lot of widespread applications, in marine, automobile, and aircraft industries. However, the manufacturing techniques have some difficulties such as the high cost of equipment and tooling [2].

Many research works have dealt with the preparation and properties of metal matrix composites (MMCs), and a review of some of them is summarized herein. The wear behavior and lubrication of two metal matrix composites were examined by Caracostas, Constantinos A. et al. [3]. The authors evaluated the wear behavior of two MMCs manufactured using the 2024 T4-aluminum alloy as the matrix and titanium diboride ($TiB_2$)

particles as reinforcement while they were in sliding and rolling contact with 52,100 steel and hardened pearlitic nodular cast iron. According to the investigation, while in sliding contact, the 10 vol.% 0.3 μm $TiB_2$-MMC exhibited marginally less wear than the 10 vol.% 1.3 μm MMC. The $TiB_2$ particles on the wear track prevented spalling, and the MMC suffered no subsurface damage, according to the authors' observations. In contrast, Run Geng et al. [4] studied the microstructure development of the Al-Mg-Si (6061) alloy after casting, cold rolling, and heat treatments with a minor addition of nanosized TiC particles (TiCp). TiCp inhibited recrystallization and grain development, refined grain structure, and enhanced tensile strength without decreasing ductility. The 1.0 wt.% TiCp/6061 composite had 330 MPa and 275 MPa in the T6 state and 438 MPa and 426 MPa in the as-rolled state, 42 MPa and 38 MPa, and 41 MPa and 52 MPa, higher than the matrix 6061 alloy, respectively. The researchers concluded that adding TiCp leads to grain refinement and increase in both ultimate tensile strength and yield. They stated that this is due to stress dislocation buildup and precipitate strengthening. Run Geng et al. [5] investigated the corrosion behavior of Al-Mg-Si alloys when adding nanosized TiC and $TiB_2$ particles (TiCp, TiB2p) and concluded that the corrosion behavior of Al-Mg-Si alloys is improved by adding TiCp and worsened by adding $TiB_2$p. This is due to grain refinement resulting from adding TiCp, whereas $TiB_2$p was segregated at grain boundaries and sped up corrosion. Furthermore, Al-Mg-Si metal matrix composites were examined by D. Mummoorthi et al. [4] to study the effects of a stir-cast reinforced alloy with 5 wt.% $Fe_2O_3$ with 2%, 4%, and 6% weight of $B_4C$. Al 6061 composites were examined for physical and mechanical parameters such density, hardness, impact strength, ultimate tensile and compressive strength, and microstructure with varied wt.% of reinforcement using optical microscopy. An SEM study showed muscular bonding between the matrix and reinforcements in freshly produced composites. They arrived at the conclusion that $Fe_2O_3$- and $B_4C$-reinforced Al6061 show improved mechanical properties. Hence, the percentage of $Fe_2O_3$ and $B_4C$ particles increased microstructure uniformity and tensile strength of metal composites without decreasing elongation. On the other hand, Mummoorthi Duraipandian [6] examined aluminum Al6061 (Al-Mg-Si) alloy, 5 wt.% $Fe_2O_3$, and 2 wt.%, 4 wt.%, and 6 wt.% $B_4C$-reinforced matrix composites. The researcher adopted stir-casting to produce the alloy. The researcher found that the mechanical characteristics of composites improved with weight %. He also examined the corrosion resistance of $Fe_2O_3$- and $B_4C$-reinforced Al6061 metal matrix composites. Materials immersed in 3.5% brine solution were tested for corrosion resistance using different electrical test parameters. SEM showed composite corrosion before and after preparation. The study found that increasing reinforced composite wt.% increases corrosion resistance. In addition, the manufacturing process was investigated by Sunghak Lee et al. [7] as they examined the microstructure of squeeze-cast and permanent mold re-cast A356 Al-SiC particle composites. These composites were toughened by notched fractures to establish their critical fracture parameters using stress-modified critical-strain criteria. The composite microstructure has continuous networks of densely populated SiC and eutectic Si particles in intercellular zones. MMC wear behavior was also investigated by R. Auras and C. Shvezov [8], who they studied five zinc–aluminum (ZA)-based alloys with silicon, copper, and 8 and 16% volume of reinforcing silicon carbide (SiC) particles. SEM, EDXA, and X-ray diffraction were utilized to study the alloying characteristics, wear surface, and wear debris of cast samples. Pin-on-disc wear was adopted in both dry and lubricated forms, and 29.43 N (3 kg), 49.05 N (5 kg), 78.48 N (8 kg), and 250 rpm (2 m/s) were used. The researchers observed that SiC particles improved the matrix alloy wear, and the ZA alloy wear rate was non-linear based on the test load. Nonreinforced alloys lost material in dry conditions. They also observed that local plastic deformation and element transfer occurred in nonreinforced alloys. In another study by Kumar, G.B.Veeresh et al. [9] also studied physical, mechanical, and tribological characteristics of the Al 6063 alloy reinforced by silicon nitride powder via stir casting. Matrix reinforcement varied from 0 to 10% in 2-wt% increments. The authors found out that the reinforcing percentage greatly increased hardness and density and that the composite wore well. Since Al 6063 is soft, this work generated Al 6063–silicon nitride

MMC with superior mechanical and tribological properties. The researchers used SEM to examine produced composites before and after wear test morphology and concluded that adding silicon nitride highly increases the wear resistance of the Al 6063-based MMC.

K.R.Padmavathia and Dr. R.Ramakrishnan [10] used stir casting after melting aluminum alloy in an electric furnace to study the wear and friction of Al 6061 (Al-Si-Mg) with varying percentage volumes of multiwall carbon nanotube and silicon carbide reinforcement. MWCNT and SiC, warmed at 620 °C, were mixed into the molten metal at 750 °C. To evenly incorporate SiC and MWCNT particles into the Al 6061 matrix, a twin-blade mild steel impeller stirred them at 450 rpm for 5 min. Nano SiC and MWCNT particles were used as reinforcement. The studies employed 0.5% and 1.0% MWCNTs and 15% SiC. The pin-on-disc apparatus was used to examine specimen wear. The researchers observed that under mild wear, the composite outperformed aluminum; meanwhile, under severe wear conditions, the composite showed a higher wear rate and friction coefficient, and it was found that the friction and wear behavior of the Al-SiC-MWCNT composite was greatly influenced by the applied load and that CNTs could negatively affect aluminum alloy wear resistance beyond a critical load, and reinforcement increased composite hardness. M.Vamsi Krishnaa and Anthony M. Xavior [11] also explored the Al6061-SiC/graphite hybrid composite's mechanical properties. Their study employed Al 6061matrix, 37 μm silicon carbide, and 1 μm graphite reinforcement. Stir-casting composites employed 5–15% reinforcement in 5wt.% increments. The authors used an electric resistance furnace to heat the aluminum alloy at 450–800 °C for 2 h before melting. Reinforcing particles (SiC) and graphite particles were heated at 600 °C and 1100 °C for 2 h to increase wetness and remove adsorbed hydroxide and other gases. Heating the matrix over 750 °C melts the metal entirely. A semi-solid melt was cooled between liquidus and solidus points. Prior to adding the particles, magnesium powder was added to melt to maintain the wettability. Then, they added the preheated particles in three separate steps and then mixed them for 10–20 min at 200–400 rpm. The researchers also warmed molds at 250–350 °C for 2 h before pouring the melt, and then the 730–800 °C melt was poured into the mold. It was observed that the composite microphotographs demonstrated particle dispersion with few clusters. They also found that composite densities were below the theoretical values and concluded that dispersed graphite and SiC in Al6061 alloy strengthened composites. SEM scans indicated void-free matrix-reinforcing particle dispersion. In addition, Oyewusi Elijah Oyedeji et al. [12] studied aluminum metal matrix composite materials to determine the best composition of Al-Mg-Si alloy reinforced with palm kernel shell ash (PKSA) particles in terms of static and dynamic characteristics. The composites were made from powder. The SEM result indicated that recrystallization during ball milling of mixed powders resulted in increased dislocation density and tougher phases in the PKSA, which enhanced their thermal characteristics. The $C_4$ sample (6:94 wt.%) had optimum characteristics since the modulus of rupture of the generated Al-Mg-Si-PKSA matrix composites was raised. Based on dynamic mechanical thermal analysis (DMTA), the developed composites' storage modulus, loss modulus, and damping factor were found. Frequency and amplitude increased Tg. This study also found that the use of 6 wt.% PKSA on Al-Mg-Si powder is ideal for aeronautical research. Moreover, Madeva Nagral et al. [13] examined the effects of micron-sized graphite addition on the Al6061 alloy's microstructure and tensile failure. The two-step stir casting approach improved the wettability of Al6061 alloy metal matrix composites enhanced with 6, 9, and 12 weight percentages of graphite particles. Microstructural, density, hardness, and tensile characteristics were tested on synthetic composites. SEM, EDS, and XRD patterns characterized samples' microstructures. Graphite reinforcing lowered metal composite density and hardness while increasing ultimate tensile strength and yield strength. Reinforcement increased the Al6061 alloy composite elongation.

As shown above, most MMC composites are based on pure metal filled with additives. Few studies deal with the segregation effect caused by aluminum alloy; moreover, Al-Si-Mg alloy is the matrix material. Regarding our hypothesis, the Al-Si-Mg alloy is strengthened by ceramic additives [14] and enhances the compressive strength, but the strengthening mechanism varies with the position of the cast in the casting mold. Segregation effects were major factors in the variation in mechanical properties [15], which also reduced the micro-homogeny distribution of alloy elements through the cast. In addition to the molding effect [16], therefore, the novelty, which needs more study, was the effect of stir casting on the segregation of the alloy, as well as the heat treatment during casting, which saves effort and time for progressive sequences and operation. Hence, the present study has three main topics of investigation: (1) the fabrication technique using a special preparation stir-casting process, (2) the effect of mold types and different cooling rates, and (3) the tribological properties of MMC based on Al-Si-Mg/SiC, $Al_2O_3$ composites.

The paper is structured as follows: in the first section, the fabrication technique is explained, followed by the aspects of thermal analysis and cooling rate; in the second section, the mechanical and tribological tests are outlined; in the third section, the results and discussion are presented; and in the last section, the conclusions and remarkable concepts are summarized.

## 2. Material and Methodologies

The material was aluminum alloy 6061, which we sourced from the Aluminum Company of Egypt, i.e., Naga Hammadi, Egypt. This type of alloy is characterized by its high strength, good weldability, and high wear resistance, due to its particular chemical compositions, which are listed in Ref. [17]. It was found that the two main constituents of the alloying element of this aluminum alloy were magnesium, with a content of (0.8–1.2) wt.%, and silicon, with a content of (0.4–0.8) wt.%. Aluminum is distinguished by high fracture toughness [18]. This observable composition was the main reason for the attractive properties of this alloy and can therefore be described as such [19–21].

### 2.1. Stir Casting Process

The aluminum segments were placed in a small ceramic crucible and then placed in an electric furnace with a maximum heating temperature of 1500 °C. The metal was heated to 650 °C. The molten metal was kept at this temperature for 2 h to achieve sufficient homogeneity and solubility [22]. This temperature was above the melting temperature of 580 °C [23]. This temperature was chosen to allow enough time to mix the ceramic additives— alumina $Al_2O_3$ and silicon carbide SiCp particles (supplied by El-Gomhouria for Medicines and Medical Supplies, Cairo, Egypt) (Swith a size of almost 40–60 µm—based on heat and mass transfer criteria [24]. The crucible was then removed from the oven and placed in the chamber of the electric mixer (see Figure 1), and the agitator was switched on while the ceramic additives were poured in. The electric mixer was made so that the metal did not splash out during the mixing and stirring process. To prevent a large temperature loss, the ceramic fan was preheated to the same temperature, and the ceramic additives were added to the furnaces to achieve the same temperature of the molten metal. The stirring process took less than 30 s, and then the crucible was poured into the molds. There were two types of molds: H13 Steel alloy of 24.3 W/m.K thermal conductivity and of 0.460 J/g-°C specific heat [25] (Figure 2a) and synthetic graphite (petroleum coke, needle coke, coal pitch, etc.) of 121.1 W/m.k thermal conductivity [26] and of 1.732 J/g °C specific heat [27] (Figure 2b). The castings produced were like the aluminum alloy without any additives (0%), while the $Al_2O_3$ and SiC particles were added at 0.5 wt.%, 1 wt.%, 2 wt.%, 3 wt.%, 4 wt.%, and 8 wt.%. It is known that these ceramic additives do not dissolve in the aluminum matrix due to the large differences between melting temperature and density; therefore, the cast product can be called a metal matrix composite [14,28,29]. The cast products had a cylindrical shape of 238 mm × 15 mm for casting in a steel mold and 200 mm × 22.5 mm for casting in a graphite mold.

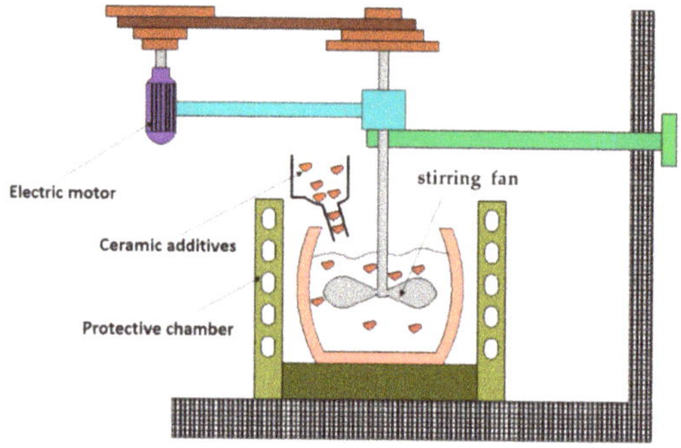


**Figure 1.** Stirring process with protective chamber.

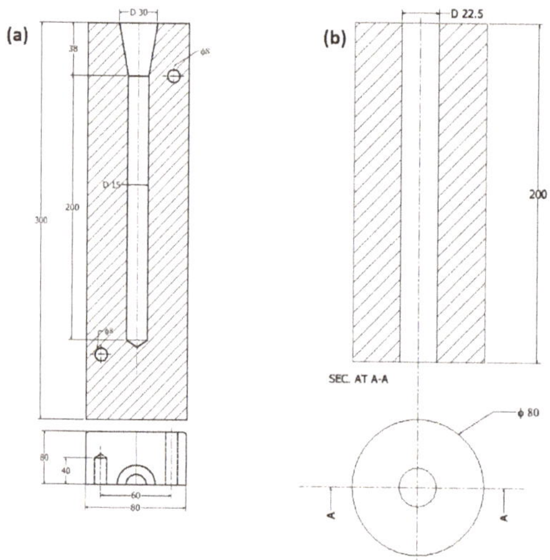


**Figure 2.** Casting molds: (**a**) steel, (**b**) graphite.

*2.2. Heat Loss Analysis*

The mechanism of heat transfer during casting depends on the solidification process, the cast metal, and the thermal radiation [30]. These dynamics can be explained by the cooling curve shown in Figure 3. The heat loss during the casting process up to $T_l$, known as the cooling temperature of the liquid or the beginning of solidification, is mainly due to the turbulent flow in the molten metal to compensate for the longer time needed for mixing the additives and the stirring process; this point is smaller than $T_{ms}$. The rate of solidification in this range is linear, and the amount of heat is minimal. Solidification begins on contact with the molds at $T_l$, then continues gradually until it solidifies completely at temperature $T_s$. The castings are left in the molds until they have cooled completely to room temperature Tr. This normalizes the grain size of the product in the molds, and the boundary conditions are as homogeneous as possible. The solidification process progresses over time. The time at which the temperature of the molds rises significantly to the maximum temperature is the same as when the molten metal reaches the liquidus point. Therefore, a thin layer forms over the outer casting surface, and then the metal shrinks and an air gap is interrupted between the mold and the metal.

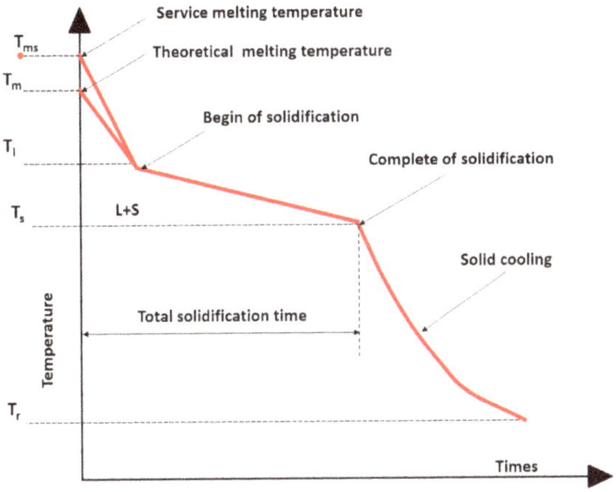


**Figure 3.** Theoretical solidification curve of the alloy.

Thermal Analysis

Casting of aluminum involved moving the molten metal from the furnace to the mold through a refractory casting crucible; during this operation, the heat was lost into the atmosphere, the crucible during the stirring operation, and even through the casting mold [31]. The heat loss caused by contact with the crucible and the mold can be calculated using Fourier's law in one dimension:

$$\frac{\partial T}{\partial t} = \alpha \frac{\partial^2 T}{\partial y^2} \tag{1}$$

where T is the temperature in metal 650 °C and y isdisplacement through the mold depth. Using the boundary conditions on the inside of crucible and mold that are the same as the metal, therefore, we can set y = 0 and T (0, t) = T, y = $-\infty$, T ($-\infty$) = Ts, whereas the heat flux at the interface between the wall of the crucible and mold [31] is

$$\frac{\partial q(0,t)}{\partial t} = k\frac{\partial T(0,\ t)}{\partial y} \tag{2}$$

By solving Equation (1) with respect to the derivation of Equation (2), the heat flux P can be calculated in watts using Equation (3) as follows:

$$P = \sqrt{\frac{k\rho c_p}{\pi \times t}} A(T - Ts) \tag{3}$$

where k is the thermal conductivity of the ceramic refractory for the crucible or for the mold material (graphite, or steel), and $c_p$ is specific heat. A is the surface area normal to the direction of heat transfer and can be calculated as follows:

$$A = s\,(2h + b) \tag{4}$$

where s is distance per second of metal flow (m), h is the metal depth, and b is crucible or mold width. For a mass m molten metal, the energy can be expressed by rewriting Equation (3) as follows:

$$P = m \times c_p \times \Delta T \tag{5}$$

Hence, the temperature gradient or loss $\Delta T$ in transferring the molten metal from furnaces to the crucible or mold cavity is as follows:

$$\Delta T = Pm \times c_p \tag{6}$$

For the heat reduction due to radiation, the temperature loss can be calculated using the radiation as follows [31]:

$$P = \varepsilon \times \sigma \times A\left(T^4 - {T_o}^4\right) \tag{7}$$

where $\varepsilon$ is the emissivity (0.09 to 0.24) for the aluminum alloy 6061 [32] and $\sigma$ is Stefan Boltzmans constant [33,34], $5.67 \times 10^{-8}$ (W/mK). Then, the temperature loss can be calculated using Equation (8):

$$\Delta T = \frac{\varepsilon \times \sigma \times b\left(T^4 - {T_o}^4\right)}{m \times c_p} \tag{8}$$

where $\Delta T$ is now the temperature reduction per meter due to radiation.

### 2.3. Sample Perspiration

The cast products taken from the molds were cut into nominal dimensions using the Turing process. The nominal dimensions were 12 mm × 200 mm for the specimens cast in steel molds, while they were 180 mm × 20 mm for those cast in graphite molds. These specimens were cut into three equal segments along the entire length of the specimen. This technique reduced the inhomogeneity and percentage error when repeating the test, as it was the same sample and the segregation effect was low, as cited in a previous study by Hassan et al. [14,29].

### 2.4. Compression Test

The compression test was carried out according to the ASTM E9 standard [35] at room temperature. This test is used to measure the compressive strength, Young's modulus, and percentage deformation of the cast product. The test was conducted using a universal testing machine (Model WDW-100, Jinan Victory Instrument Co., Ltd., Jinan, China) [36] with a load capacity of 100 kN and a controlled speed of 5 mm/min. The load values were recorded with the load indicator, while the deformation was measured with a 0.005 mm dial gauge attached to the moving platen of the machine. The ends of the specimens were supported on two hardened parallel steel plates. These two parallel plates, conforming to the ASTM E9 standard [35], were used to ensure even distribution of the loads and to prevent tipping of the head, which can occur during the test. The test was performed dry and without lubrication between the plate and the specimen. The test was continued until the optical crack imitation appeared on the surface of the specimen. The sample height was measured before and after the test, with a 0.05 mm vernier. The test was performed on all samples with and without additives, cast in both molds. The specimens were, as mentioned above, short specimens with a high thickness ratio (h/d $\approx$ 0.8) according to the ASTM E9 standard [35].

### 2.5. Hardness Test

The Vickers hardness test was conducted in accordance with ASTM E384-10 [37]. The test was performed with a universal hardness testing machine (Zwick/ZHU 187.5, ZwickRoell, Ulm, Germany) [26]. A diamond-shaped, square, pyramid indenter was pressed into the surface under a uniform load. The load was maintained for (15) seconds to achieve a uniform condition, and then the indenter was removed. The surface of the specimen should be prepared by polishing in two steps before testing. The first step is polishing with medium-grit emery paper (1000) to remove all rough surface imperfections, and the second step is to use very fine silicon carbide emery paper (2000) to polish the surface like a mirror. Then, the surface is etched with sulphuric acid $H_2SO_4$ for about 1 min.

After that, the samples are ready for the Vickers hardness test. The specimens taken from the three previously cut positions were 5 mm × 12 mm and 5 mm × 20 for the specimens cast in steel and graphite molds, respectively.

## 2.6. Pin-on-Disc Wear Test

The wear test is performed with a pin on a disc according to the ASTM G99-05 standard [38]. The test was performed on all cast products. Figure 4 shows a general view of the wear test machine (SSITOM-012-Pin On Disc Wear Apparatus, Saini Scientific Industry, Haryana, India) [14]. It consists of a horizontal steel disc with an inner diameter of 50 mm, an outer diameter of 100 mm, and a thickness of 2 mm, which rotates and is driven by a variable-speed motor. A specimen holder connected to the loading lever and pressed against the rough counter surface is used to hold the specimen. The load is applied with a dead weight of 1600 g at a 3.5 m/min and 5 m/min linear speed at 5 min contact time. The surface of the sample was polished with 1000-grit emery paper to remove any unevenness or impurities. The initial weight was then measured with an accuracy of 0.001 g, and the sample was fixed in the play as a 12 mm diameter pin. In addition, 80-grit emery paper with a surface roughness of $1.80 \times 10^{-6}$ was glued to the opposite side of the steel disc with epoxy glue; then, the pin (the sample surface) was contacted in a dry state and not lubricated. The motor speed was 400 rpm.

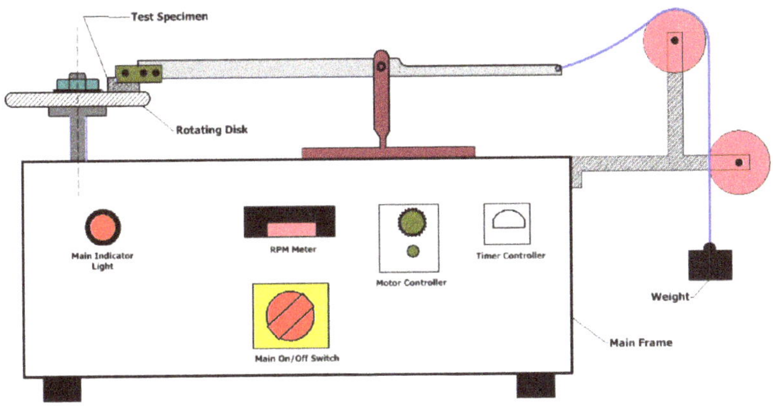


**Figure 4.** Pin-on-disc machine.

# 3. Results and Discussion

## 3.1. Compressive Strength

The average compressive strength as a function of deformation stress is shown in Figures 5 and 6. It was found that the deformation strain increases with increasing additions for $Al_2O_3$ cast in graphite mold (Figure 5a). This was true for specimens with 1, 4, and 8 wt.% $Al_2O_3$. In addition, these three specimens had a higher compressive strength. The Young's moduli are close to each other, the curve is almost smooth, and no catastrophic failure occurred. The strength reached almost the maximum values of over 500 MPa for the specimens without $Al_2O_3$, while it decreased with increasing $Al_2O_3$ content in the steel mold (see Figure 5b), which can be attributed to deformation. The values of the Young's moduli were almost the same. This is because the alloy cast in a steel mold has a lower specific heat and a higher emission coefficient ε, so the heat transfer is higher and faster for steel castings than for samples cast in a graphite mold. The higher cooling rate resulted in refinement and smaller grain size [16], while in graphite molds and ceramic molds, which are characterized by higher specific heat and thus lower emission coefficient, heat transfer takes longer, resulting in the coarsening of the grains. A fine grain size is preferred to increase the strength and reduce voids and defects. Figure 6a shows the compressive strength of composite aluminum embedded with SiC particles casted in a graphite mold. The strength values and deformation strains for 0.5, 3, and 8 wt.% Si C were nearly the same; with the as-received aluminum alloy (without additives), the deformation strain reached a maximum of 0.5 for 8 wt.% SiC and a minimum for 2 wt.% SiC. For composite metal casted in steel mold (Figure 6b), it was found to increase in strength and deformation strain with the increase in the wt.% of SiC additives for the aluminum alloy matrix. However, the strength and deformation strain decreased for 1 wt.% and 3 wt.% Si C, due to segregation in the alloy, which cannot be eliminated completely in the casting production technique [14,29]. More concentration appears on the curve, and it was found that the curve has two sloped regions, an elastic slope followed by another slope over the plastic zone, as clearly seen in specimens of 1 wt.% and 4 wt.% $Al_2O_3$ casted in steel mold (see Figures 5b and 6b). Tables 1 and 2 list the exact values of the average of three compressive tests of Al-Mg-Si alloy composed of ceramic additives casted in graphite and steel molds, respectively. In Table 2, the specimens gave high average compressive values for 1 wt.% $Al_2O_3$ of 530.37 MPa with a standard Deviation SDV 1.06 and a maximization as 55.45 MPa with SDV 29.12 MPa at 4 wt.% $Al_2O_3$ when casted in the graphite mold. In addition, it was 525.98 MPa with SDV 3.37 MPa for 1 wt.% SiC and 545.92 MPa with SDV 3.7 MPa at 4 wt.% SiC when also casted in graphite molds. The lowest percentage error was 0.61 and 0.9 for specimens with 1 wt.% $Al_2O_3$ and 3 wt.% SiC, which indicates the stability of the test procedure and its repeatability. The highest value of the height reduction was 54.4% for the specimen containing 4 wt.% SiC particles; therefore, the ductility for this specimen and the deformation ability was increased and enhanced, which is the reason behind the higher compressive properties with a composite containing Si C and a symmetry between the alloy internal element over 0.5 wt.% Si (see Figure 7a,b). Table 2 lists the effect of the steel mold on the casting compressive strength; it was found that the maximum compressive strength was 623.53 MPa with an SDV of 24.99 MPa for 8 wt.% $Al_2O_3$ particles. Then, asecond degree was a specimen filled with 4 wt.% Al2O3 as 574.77 MPa with an SDV of 19.08 MPa, whereas for SiC particles, the reduction in strength in terms of the maximum values walso at 4 wt.% SiC with 539.46 MPa and an SDV of 16.3 MPa, whereas, in the second grade, the 8 wt.% SiC had an average value of compressive strength of 537.65 and an SDV of 3.59 MPa. The higher ductility was achieved with a 4 wt.% of SiC of 59%. Although the casting in steel mold was given higher values of compressive properties, the stability of the tests was smaller than that of casting in the graphite mold, because the % error of the specimen casted in steel had a higher value, and the minimum value was 1.04 for 4 wt.% $Al_2O_3$. Then, in the second grade, the value was 1.49 for 3 wt.% SiC particles (see Figure 7a,b). Figure 8 shows an optical image to illustrate the distribution of the particle through the Al-Si-Mg alloy matrix. It was observed that there was a region with a higher concentration than other regions. The surface patterns wer finer for the specimen casted in the steel mold, as shown in Figure 8a,c, whereas some coarsening was observed in the specimen casted in graphite molds, as shown in Figure 8b,d. The load was transmitted between the MMC by the interfacial region between the Al-Si-Mg alloy and the SiC and $Al_2O_3$ additives [39,40]. Therefore, the increase in compressive strength with the additives occurs because these additives function as a secondary feature in the matrix phase, which resists the progressive deformation in the metal matrix [4,40]. While the separation at the interface between the Al-Si-Mg alloy and theceramic additives (SiC and $Al_2O_3$) was the major failure mechanism in the recasting methods [7], the failure mechanism related to the alloy without the additive depends on the Si element, which was homogeneously distributed in the aluminum matrix; therefore, it cleaved to form microcracks [35]. Therefore, the separation at the interfaces formed ductile cracks ahead of microcracks and then interconnected by localized shear bands deformed in the aluminum matrix. These results were similar to those of Nikhil et al. [41], where the graphite was used to strengthen the aluminum matrix composite. It was reported that the strength enhancement due to the reinforced phase acts as if they are a load barrier [7,13].

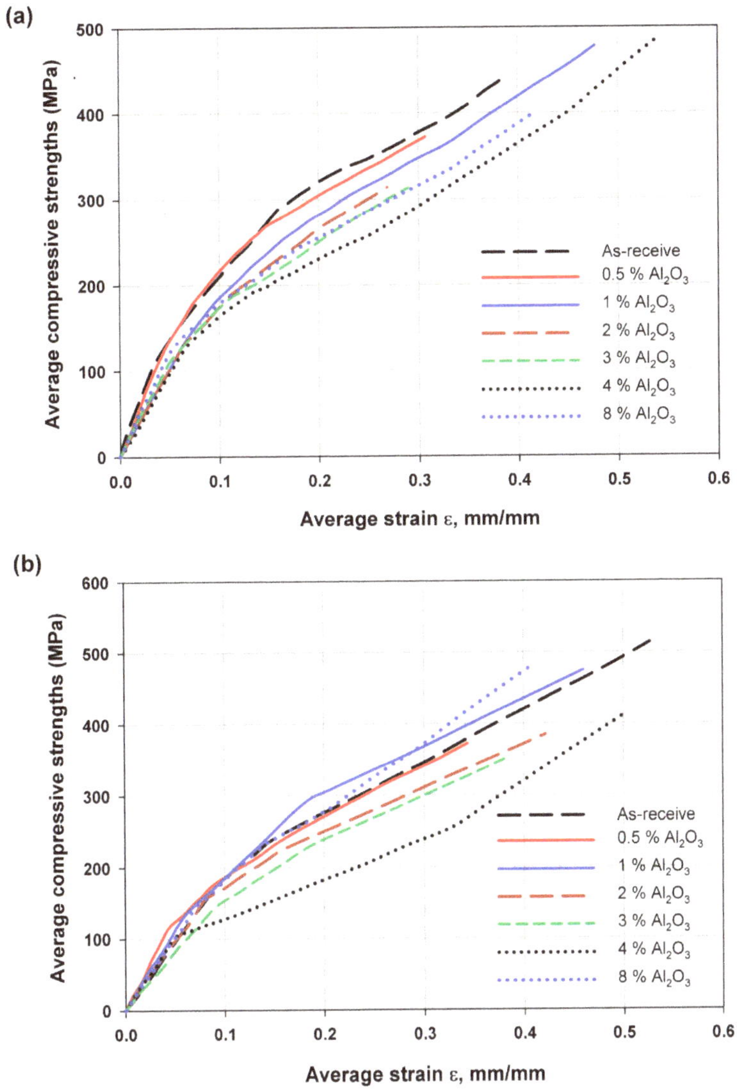


**Figure 5.** Average stress and strain relation for the Al-Si-Mg alloy filled with $Al_2O_3$ in (**a**) graphite mold, (**b**) a steel mold.

**Table 1.** Compression test results for the ($Al_2O_3$ and SiC)/Al-Si-Mg composite casted in a graphite mold.

| wt.% Additives | Compressive Strength σ, MPa | | | | | | % Height Reduction | |
|---|---|---|---|---|---|---|---|---|
| | $Al_2O_3$ | SDV | % Error | SiC | SDV | % Error | $Al_2O_3$ | SiC |
| Without additives | 504.88 | 9.06 | 5.23 | 504.88 | 5.23 | 5.23 | 39 | 39 |
| 0.5 wt.% | 378.87 | 27.07 | 15.62 | 411.72 | 26.32 | 15.19 | 30.7 | 34.8 |
| 1 wt.% | 530.37 | 1.06 | 0.61 | 545.92 | 3.37 | 1.95 | 47.7 | 39 |
| 2 wt.% | 409.30 | 17.27 | 9.97 | 373.36 | 14.54 | 8.39 | 26.8 | 26.8 |

**Table 1.** *Cont.*

| wt.% Additives | Compressive Strength σ, MPa | | | | | | % Height Reduction | |
|---|---|---|---|---|---|---|---|---|
| | $Al_2O_3$ | SDV | % Error | SiC | SDV | % Error | $Al_2O_3$ | SiC |
| 3 wt.% | 387.69 | 18.3 | 10.50 | 353.61 | 0.9 | 0.52 | 29 | 19.7 |
| 4 wt.% | 555.43 | 29.12 | 16.81 | 525.98 | 3.7 | 2.13 | 54.4 | 49.7 |
| 8 wt.% | 491.85 | 26.4 | 15.29 | 519.61 | 12.3 | 7.10 | 41.9 | 43.1 |

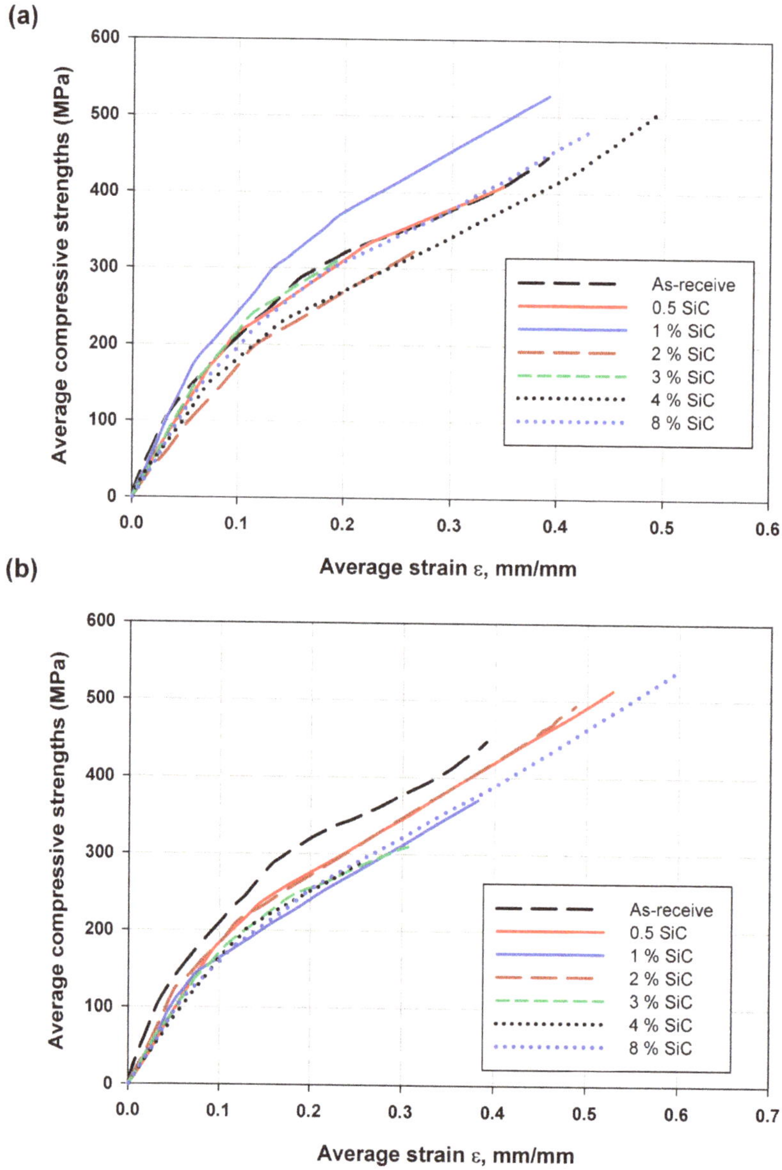


**Figure 6.** Average stress and strain relation for the Al-Si-Mg alloy filled with SiC in (**a**) graphite mold, (**b**) a steel mold.

**Table 2.** Compression test results for ($Al_2O_3$ and SiC)/Al-Si-Mg casted in a steel mold.

| wt.% Additives | Compressive Strength σ, MPa | | | | | | % Height Reduction | |
|---|---|---|---|---|---|---|---|---|
| | $Al_2O_3$ | SDV | % Error | SiC | SDV | % Error | $Al_2O_3$ | SiC |
| Without additives | 513.3 | 30.07 | 17.36 | 513.31 | 30.07 | 17.36 | 52.6 | 52.6 |
| 0.5 wt.% | 396.85 | 32.64 | 18.82 | 370.57 | 22.31 | 12.88 | 34.3 | 38.1 |
| 1 wt.% | 513.29 | 5.91 | 3.42 | 519.12 | 25.73 | 14.86 | 46 | 48.6 |
| 2 wt.% | 384.73 | 1.81 | 1.04 | 343.66 | 9.18 | 5.30 | 42 | 30.5 |
| 3 wt.% | 363.29 | 13.56 | 7.80 | 343.50 | 2.58 | 1.49 | 38 | 25.8 |
| 4 wt.% | 574.77 | 19.08 | 11.02 | 539.46 | 16.93 | 9.77 | 50 | 59.8 |
| 8 wt.% | 623.53 | 24.99 | 14.42 | 537.65 | 3.59 | 2.07 | 41 | 34.4 |

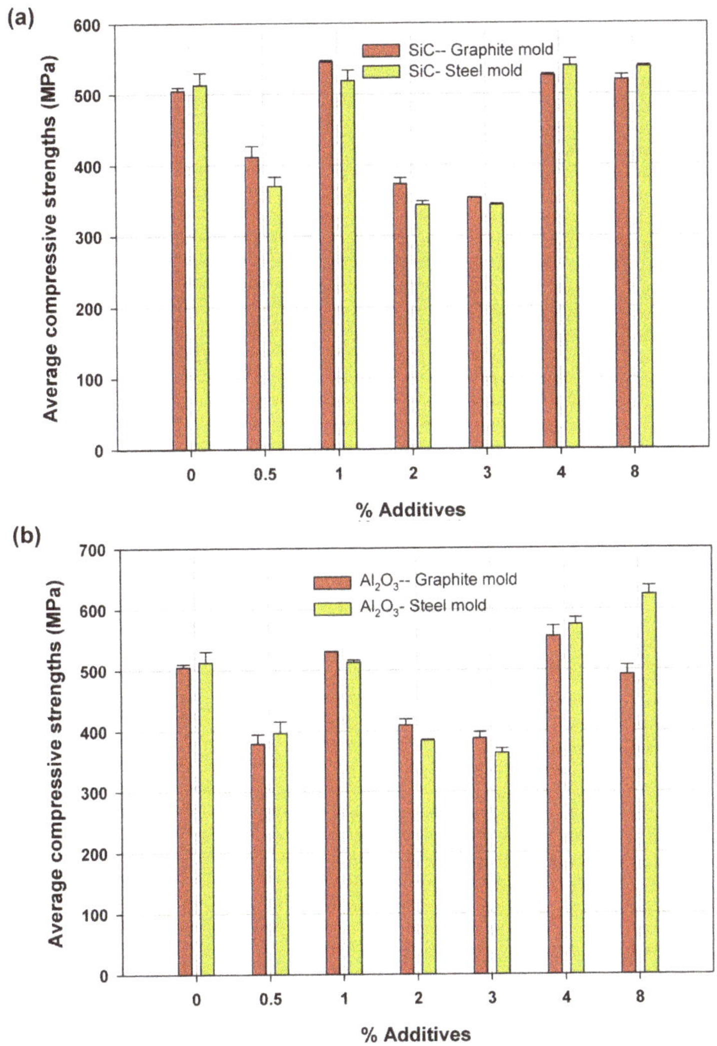


**Figure 7.** Average stress variation for the Al-Si-Mg alloy filled with (**a**) SiC and (**b**) $Al_2O_3$ particles.

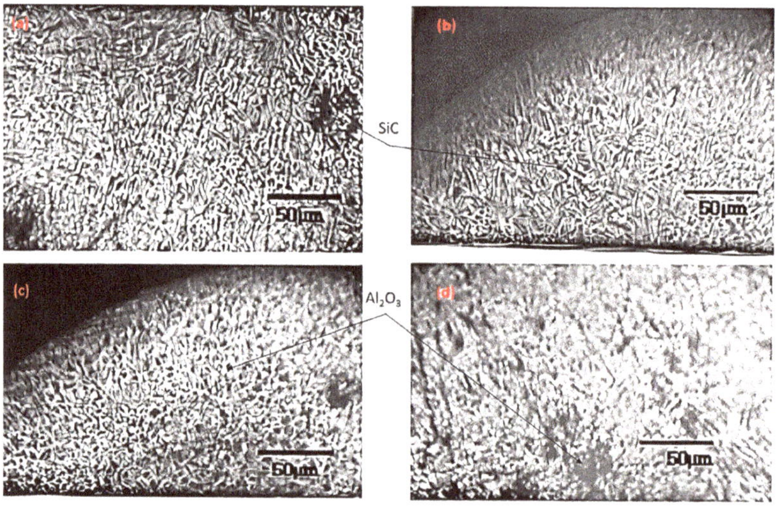


**Figure 8.** Optical microscopy images: (**a**) steel mold, (**b**) graphite mold for 1 wt.% SiC particles, and (**c**) steel, (**d**) graphite mold for 8 wt.% $Al_2O_3$ particles.

### 3.2. Hardness Test

The change in Vickers hardness at different $Al_2O_3$ weight percentages is shown in Figure 9a. It was found that there are two factors. The first factor, Vickers hardness, was higher for 0 wt.%, 0.5 wt.%, 1 wt.% and 2 wt.%, and 1 wt.% and 2 wt.% $Al_2O_3$ cast in steel molds, while for the second issue, for 3 wt.%, 4 wt.% and 8 wt.% $Al_2O_3$, the trend was towards higher hardness. For the % $Al_2O_3$, the trend changed: the hardness was higher for $Al_2O_3$ cast in a graphite mold. This is because a cooling rate in the graphite mold is allowed due to the higher specific heat of graphite and the increasing amount of alumina, which has a lower density and takes more time to segregate through the molten metal, so it is deposited more on the surface, whereas with the small amount, it is not saturated over the entire surface of the samples [21] (see Figure 8a). The maximum Vickers hardness was 62 kg/mm$^2$ with an SDV of 0.81 kg/mm$^2$ and a % error of 0.47 for 8 wt.% $Al_2O_3$ cast in a graphite mold, and 59.5 kg/mm$^2$ with an SDV of 9.25 kg/mm$^2$ and a standard % error of 5.33 for 1 wt.% $Al_2O_3$ cast in a steel mold. In general, the standard % errors were low; therefore, the test stability was acceptable and good [42]. On the other hand, the effect of SiC particles on Vickers hardness is shown with the mold in Figure 8b. Two observations were made: first, the increase in SiC particles increased the Vickers hardness for 1 wt.% to 8 wt.%, and second, the hardness in the graphite mold was higher than that in the steel mold for all wt.%. This is due to the homogeneity of the Si addition in the Al-Si-Mg alloy, but also due to the graphite mold with SiC, as the carbon combined with graphite increased the grain size due to the low cooling rate. The maximum Vickers hardness was 66 kg/mm$^2$ for a composite with 8 wt.% SiC cast in a graphite mold with an SDV of 5.9 kg/mm$^2$ and a minimum standard error of 3.4. The maximum hardness was 56.33 kg/mm$^2$ for a composite with 3 wt.% SiC cast in a steel mold with an SDV of 4.49 kg/mm$^2$ and a lower standard error of 2.59, again indicating the reliability of the test procedures. The ceramic additives give some shielding to the aluminum matrix from plastic deformation; therefore, the strain energy stored increased, and the hardness improved [43,44]. However, the specimens casted in graphite with $Al_2O_3$ reinforcement decreased as in Seah et al. [45]. This could be attributed to the graphite molds, permitting graphite elements to diffuse through the aluminum matrix and therefore causing the material ductility to increase [46]; therefore, it was easily deformed with the hardness indentor [13], whereas with the SiC particles, it was a little higher.

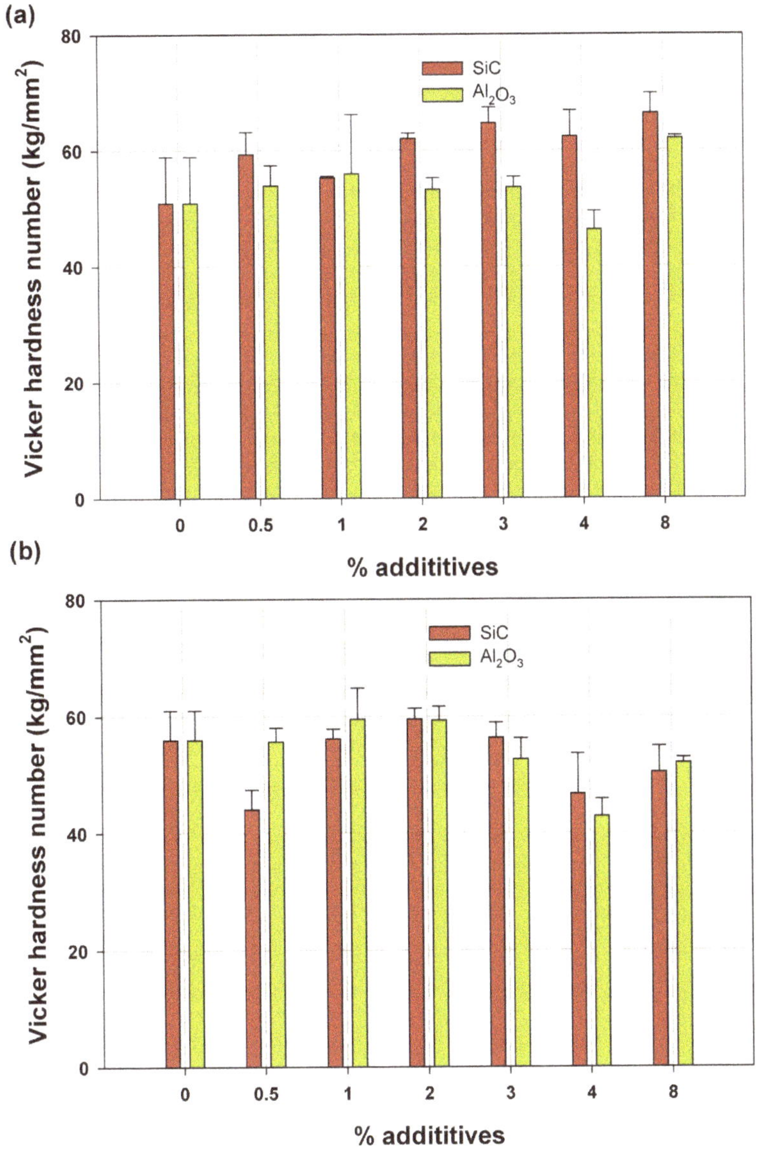


**Figure 9.** Vicker's hardness variation with different ceramic additives: (**a**) graphite mold, (**b**) steel mold.

### 3.3. Wear Properties

First of all, it is well-known that wear tests involve a variety of individual concepts, and many factors can influence material performance [47]. Therefore, various trends were observed. In general, it was found that most percentage standard errors were high because many factors affect the wear test, such as the type of interfaces, surface finish of the material, types of material, machine speed, vibration-specific heat, and thermal conductivity of the material. It was found that, in general, casting in a steel mold at a sliding speed of 3.5 m/min (see Figure 10) and 5 m/min (see Figure 11) resulted in a higher wear rate than casting in a graphite mold. This was attributed to the fact that when casting in a graphite mold, a graphite element diffuses through the Al alloy matrix, as high temperatures and good contact favor the diffusion and wettability of the aluminum [48], so graphite acts as a solid lubricant that reduces the effects of wear. In addition, the low cooling rate in graphite molds was the main reason for the coarsening of the grain size of the Al-Si-Mg alloy so that it absorbed some of the shear stress such as dumpers, resulting in a reduction in the wear loss. In contrast, when casting in steel molds, the heat loss is higher due to the higher heat transfer from radiation and the lower specific heat of the steel mold. This higher cooling rate resulted in grain refinement, especially near the mold wall [16]. In Figures 10a and 11a,b, it can be found that the 2 wt.% $Al_2O_3$ and SiC almost gave a maximum average weight loss. It was 506.66 mg with an SDV of 123.5 mg and a minimum standard error of 71% for the Al-Si-Mg alloy cast in a graphite mold filled with $Al_2O_3$ at a sliding speed of 3.5 m/min (Figure 10a), while it was 436 mg with an SDV of 200 mg and an error of 115.5% error for the alloy filled with $Al_2O_3$ also cast in a graphite mold but at a sliding speed of 5 m/min (see Figure 11a). Furthermore, the SiC at 2 wt.% shows the same trend, as the average wear loss was 776.6 mg with an SDV of 212 mg and 122% error, even when cast into a graphite mold. In contrast, the addition of 3 wt.% SiC to the Al-Si-Mg alloy increased the weight loss to a maximum value of 276 mg with an SDV of 160 mg and 92% error when also cast in a graphite mold and at a sliding speed of 3.5 m/min. It was also observed that most Al-Si-Mg alloys filled with SiC particles had a weight loss of less than 200 mg, as SiC reacts with the aluminum according to the reaction shown below, resulting in wettability of the aluminum and an increase in the amount of Si in the aluminum matrix itself [48], while Al4C3 increased the amount of carbon, which acts as a solid lubricant.

$$3SiC + 4Al \quad \rightarrow \quad Al_4C_3 + 3[Si] \tag{9}$$

It was observed generally that the Al-Si-Mg alloy matrix gave higher wear resistance, which was due to the strong bonding between the ceramic additives and the aluminum matrix, which led to less third-body wear during the wear process, as reported in [3,49]. The same results were reported for ceramic particulates inserted through an aluminum matrix carrying most of the shear load of wear action; hence, it effectively protects the matrix alloy matrix from extreme wear loss. It was also similar to that reported with other ceramic additives such as $Fe_2O_3$ and $B_4C$, where the wear resistance improves and increases with the ceramic additives. As cited before, the SiC particles and $Al_2O_3$ strengthen the aluminum matrix failure due to the separation at the interfaces, which needed larger shear forces due to homogeneity and higher dependability. On the other hand, the wear loss mechanism may be due to the oxide layer formed around the ceramic additives at the interfaces, which may lead to the formation of MgO, which weakens the bonding, as it was not coherent with the matrix phase [50]. Increasing wear loss in the $Al_2O_3$ was observed due to a tribo-chemical reaction between alumina particulates and the steel of the machine disc [3], similarly to the study performed by Brown et al. using X-ray diffraction of the wear aspartates generated during wear action [51].

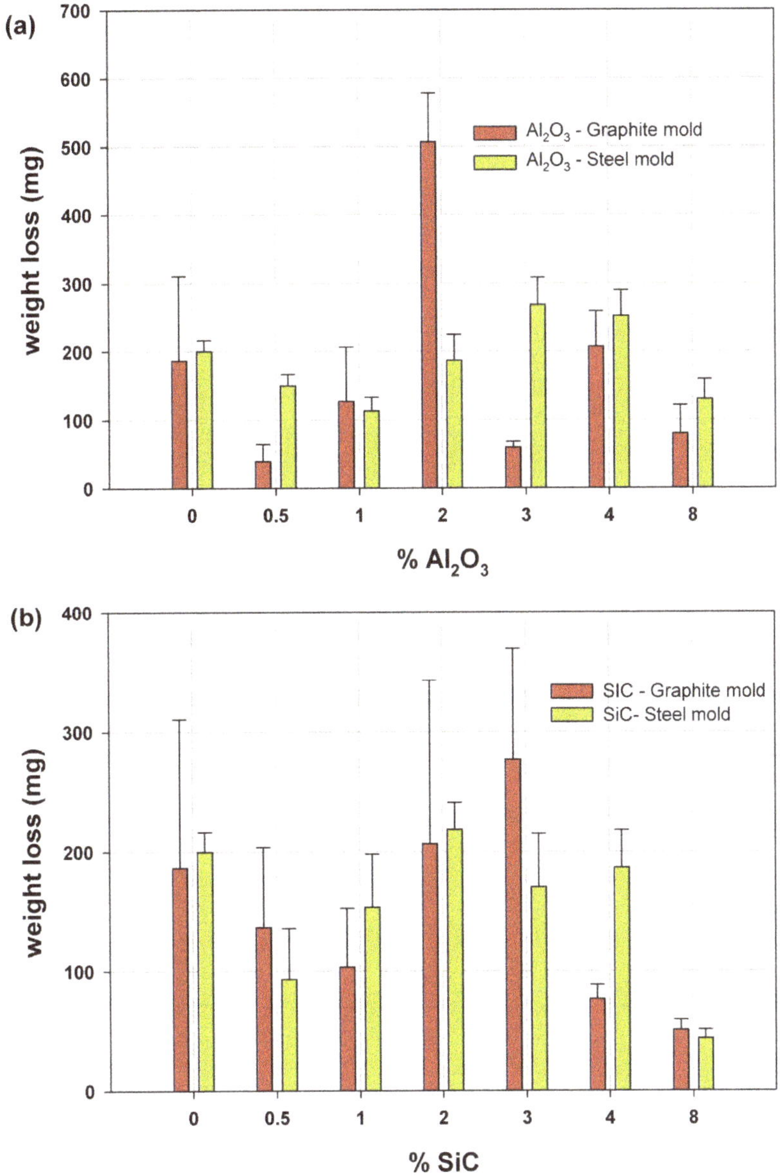


**Figure 10.** Average weight loss variation at 3.5 m/min velocity for Al-Si-Mg alloy filled with (**a**) $Al_2O_3$, (**b**) SiC.

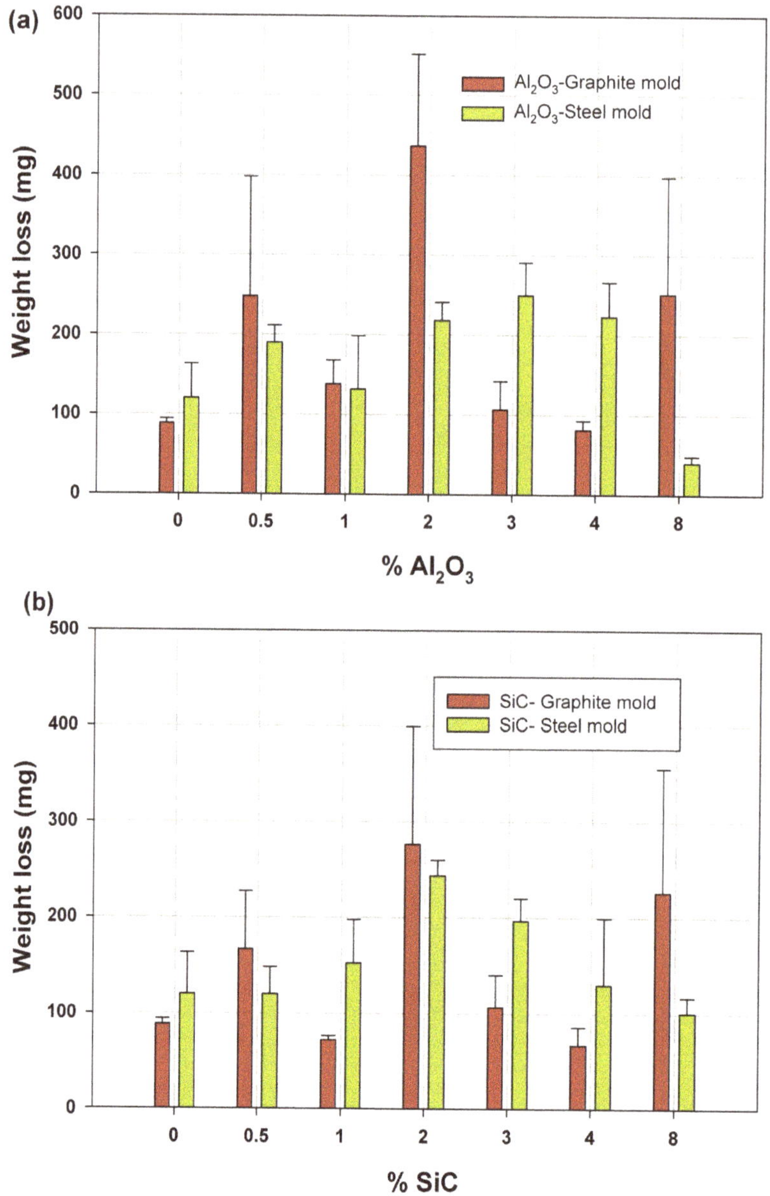


**Figure 11.** Average weight loss variation at 5 m/min velocity for Al-Si-Mg alloy filled with (**a**) $Al_2O_3$, (**b**) SiC.

## 4. Conclusions

Nowadays, metal matrix composites are setting the standard in the majority of industrial sectors. Due to its strength and high corrosion and wear resistance, the Al-Si-Mg alloy has been a competitive and alternative material in the aerospace and automotive industries. With the notably inexpensive stir-casting method, an Al-Si-Mg alloy composite material with various weight percentages of SiC and $Al_2O_3$ particles was prepared. The aluminum alloy loaded with 8 percent $Al_2O_3$ by weight showed a maximum compressive strength of

623.53 MPa. The compressive strength of the alloy with the added $Al_2O_3$ particles cast in a steel mold was 21% higher than that of the Al-Si-Mg alloy. Additionally, the aluminum alloy loaded with 8 percent SiC particles by weight and cast in a graphite mold increased the Vickers hardness by 29% to a maximum value of 66 kg/mm$^2$. However, casting in graphite mold with SiC particles was favored and is recommended owing to the involvement of Si from the SiC, graphite production from the graphite mold, and segregation of the alloy itself. Additionally, wear resistance was typically boosted by the inclusion of particles. When cast in a graphite mold, the particles in the Al-Si-Mg alloy were more evenly dispersed. On the other hand, because of the rapid cooling, the compressive strength rose when the material was cast in a steel mold. The present hypotheses were achieed through the ability to apply heat treatments to casting products during the casting process, which reduced time and expenses. An MMC with ceramic additives may be formed from an Al-Si-Mg alloy using the suggested stir-casting sequence method with a lower segregation percentage. Additionally, it had a limitation whereby MgO, the primary cause of material embrittlement, formed as a result of the creation of an oxidation layer surrounding the ceramic additions that reacted with the alloy's Mg. It is suggested that future research thoroughly examine the impact of ceramic additives at the micro-structure level using the wet wear test, SEM, and EDX analysis.

**Author Contributions:** Conceptualization, M.Y.A., M.K.H. and A.F.M.; methodology, M.Y.A. and A.F.M.; software, M.Y.A. and M.K.H.; validation, A.H.B. and M.K.H.; formal analysis, M.Y.A., M.K.H. and A.H.B.; investigation, M.Y.A., M.K.H. and A.F.M.; resources, A.H.B.; writing—original draft preparation, M.Y.A., M.K.H., B.M.F., H.M.A.E.-A. and A.F.M.; writing—review and editing, M.Y.A. and A.F.M.; supervision, M.K.H., B.M.F., H.M.A.E.-A. and M.Y.A.; project administration, M.Y.A. and A.H.B.; Editing, A.H.B. All authors have read and agreed to the published version of the manuscript.

**Funding:** This research received no external funding.

**Institutional Review Board Statement:** Not applicable.

**Informed Consent Statement:** Not applicable.

**Data Availability Statement:** The data presented in this study are available on request from the corresponding author.

**Conflicts of Interest:** The authors declare no conflict of interest.

## References

1. Biswal, S.R.; Sahoo, S. A Comparative analysis on physical and mechanical properties of aluminum composites with alo and ws reinforcement. In *Recent Advances in Mechanical Engineering*; Springer: Berlin/Heidelberg, Germany, 2023; pp. 597–603.
2. Sujatha, S. Valuation of Strength Enhancement in Heat Treated SiC Nanoparticles Reinforced 2024 Aluminum Alloy Metal Matrix Composites Using FEA. Ph.D. Thesis, College of Engineering, JNT University, Hyderabad, India, 2005.
3. Caracostas, C.A.; Chiou, W.A.; Fine, M.E.; Cheng, H.S. Tribological properties of aluminum alloy matrix TiB2 composite prepared by in situ processing. *Metall. Mater. Trans. A* **1997**, *28*, 491–502. [CrossRef]
4. Mummoorthi, D.; Rajkumar, M.; Kumar, S.G. Advancement and characterization of Al-Mg-Si alloy using reinforcing materials of Fe 2 O 3 and B 4 C composite produced by stir casting method. *J. Mech. Sci. Technol.* **2019**, *33*, 3213–3222. [CrossRef]
5. Geng, R.; Jia, S.Q.; Qiu, F.; Zhao, Q.L.; Jiang, Q.C. Effects of nanosized TiC and TiB2 particles on the corrosion behavior of Al-Mg-Si alloy. *Corros. Sci.* **2020**, *167*, 108479. [CrossRef]
6. Duraipandian, M. Effects of Fe 2 O 3 and B 4 C Addition on the Mechanical Properties and Corrosion Resistance of Al–Mg–Si Alloy in 3.5% Brine Solution. *Met. Mater. Int.* **2021**, *27*, 1506–1518. [CrossRef]
7. Lee, S.; Suh, D.; Kwon, D. Microstructure and fracture of SiC-particulate-reinforced cast A356 aluminum alloy composites. *Metall. Mater. Trans. A* **1996**, *27*, 3893–3901. [CrossRef]
8. Auras, R.; Schvezov, C. Wear behavior, microstructure, and dimensional stability of as-cast zinc-aluminum/SIC (metal matrix composites) alloys. *Metall. Mater. Trans. A* **2004**, *35*, 1579–1590. [CrossRef]
9. Kumar, G.V.; Panigrahy, P.P.; Nithika, S.; Pramod, R.; Rao, C.S. Assessment of mechanical and tribological characteristics of Silicon Nitride reinforced aluminum metal matrix composites. *Compos. Part B Eng.* **2019**, *175*, 107138. [CrossRef]
10. Padmavathi, K.; Ramakrishnan, R. Tribological behaviour of aluminium hybrid metal matrix composite. *Procedia Eng.* **2014**, *97*, 660–667. [CrossRef]
11. Krishna, M.V.; Xavior, A. An investigation on the mechanical properties of hybrid metal matrix composites. *Procedia Eng.* **2014**, *97*, 918–924. [CrossRef]

12. Oyedeji, O.E.; Dauda, M.; Yaro, S.A.; Abdulwahab, M. Characterization of Al-Mg-Si Alloy Reinforced with Optimum Palm Kernel Shell Ash (PKSA) Particle and its Consequence on the Dynamic Properties for Aerospace Application. 2021. Available online: https://assets.researchsquare.com/files/rs-795334/v1/7d9afc38-d01a-4da0-baf7-0beb6f35b473.pdf?c=1631888198 (accessed on 12 January 2023).
13. Nagaral, M.; Auradi, V.; Bharath, V.; Patil, S.; Tattimani, M.S. Effect of micro graphite particles on the microstructure and mechanical behavior of aluminium 6061 (Al-Mg-Si) alloy composites developed by novel two step casting technique. *J. Met. Mater. Miner.* **2021**, *31*, 38–45. [CrossRef]
14. Mohamed, K.; Hassan, Y.M. Abu El-Ainin H, Improvement of Al-6061 alloys mechanical properties by controlling processing parameters. *Int. J. Mech. Mechatron. Eng. IJMME-IJENS* **2012**, *12*, 14–18.
15. Seong, H.; Lopez, H.; Rohatgi, P. Microsegregation during solidification of graphitic fiber-reinforced aluminum alloys under external heat sinks. *Metall. Mater. Trans. A* **2007**, *38*, 138–149. [CrossRef]
16. Liang, G.; Ali, Y.; You, G.; Zhang, M.X. Effect of cooling rate on grain refinement of cast aluminium alloys. *Materialia* **2018**, *3*, 113–121. [CrossRef]
17. Emrani, V.; Zandi, M.; Asadollahzadeh, H. The Inhibitory Effect of Amoxicillin on Aluminum Corrosion in a Gel Electrolyte. *Int. J. Electrochem. Sci.* **2022**, *17*, 2. [CrossRef]
18. Abdellah, M.Y. Essential work of fracture assessment for thin aluminium strips using finite element analysis. *Eng. Fract. Mech.* **2017**, *179*, 190–202. [CrossRef]
19. Cheng, J.; Zhang, Z.; Dong, X.; Song, G.; Liu, L. A novel post-weld composite treatment process for improving the mechanical properties of AA 6061-T6 aluminum alloy welded joints. *J. Manuf. Process.* **2022**, *82*, 15–22. [CrossRef]
20. Gao, J.; Cao, Y.; Lu, L.; Hu, Z.; Wang, K.; Guo, F.; Yan, Y. Study on the interaction between nanosecond laser and 6061 aluminum alloy considering temperature dependence. *J. Alloys Compd.* **2022**, *892*, 162044. [CrossRef]
21. Abdellah, M.Y. Ductile Fracture and S–N Curve Simulation of a 7075-T6 Aluminum Alloy under Static and Constant Low-Cycle Fatigue. *J. Fail. Anal. Prev.* **2021**, *21*, 1476–1488. [CrossRef]
22. Anyalebechi, P.N. Analysis of the effects of alloying elements on hydrogen solubility in liquid aluminum alloys. *Scr. Metall. Et Mater.* **1995**, *33*, 1209–1216. [CrossRef]
23. Kaufman, J.G. *Introduction to Aluminum Alloys and Tempers*; ASM International: Almere, The Netherlands, 2000.
24. Middleman, S. *An introduction to Mass and Heat Transfer: Principles of Analysis and Design*; John Wiley & Sons: Hoboken, NJ, USA, 1997.
25. Material Properties Data. Available online: https://www.matweb.com/search/datasheet_print.aspx?matguid=e30d1d103816480 8a85cf7ba6aa87ef7 (accessed on 12 January 2023).
26. Available online: http://www.newmaker.com/product-1000-Hardness-Testing-Instrument-Zwick-ZHU-187,5.html (accessed on 12 January 2023).
27. Available online: https://www.matweb.com/search/DataSheet.aspx?MatGUID=3f64b985402445c0a5af911135909344 (accessed on 30 January 2023).
28. Kaczmar, J.; Pietrzak, K.; Włosiński, W. The production and application of metal matrix composite materials. *J. Mater. Process. Technol.* **2000**, *106*, 58–67. [CrossRef]
29. El-Aini, H.A.; Mohammed, Y.; Hassan, M. Effect of mold types and cooling rate on mechanical properties of Al alloy 6061 within ceramic additives. In Proceedings of the Second International Conference of Energy Engineering, ICEE-2, Aswan, Egypt, 27–29 December 2010.
30. Gowsalya, L.A.; Afshan, M. Heat transfer studies on solidification of casting process. In *Casting Processes and Modelling of Metallic Materials*; IntechOpen: London, UK, 2021.
31. Kastebo, J.; Carlberg, T. Temperature measurements and modeling of heat losses in molten metal distribution systems. In Proceedings of the Light Metals 2004: Technical Sessions Presented by the TMS Aluminum Committee at the 133rd TMS Annual Meeting, Charlotte, NC, USA, 14–18 March 2004.
32. Infrarde, E.A.Z.Z.U.; Termografije, E. Emissivity of aluminium alloy using infrared thermography technique. *Mater. Tehnol.* **2018**, *52*, 323–327.
33. Kalinin, M.; Kononogov, S. Boltzmann's constant, the energy meaning of temperature, and thermodynamic irreversibility. *Meas. Tech.* **2005**, *48*, 632–636. [CrossRef]
34. Landsberg, P.T.; De Vos, A. The Stefan-Boltzmann constant in n-dimensional space. *J. Phys. A Math. Gen.* **1989**, *22*, 1073. [CrossRef]
35. *ASTM E9-09*; Standard Rest Methods of Compression Testing of Metallic Materials at Room Temperature. ASTM International: West Conshohocken, PA, USA, 2000; pp. 98–105.
36. Available online: http://www.victorytest.com/products/wdw-100e-computerized-electromechanical-universal-testing-machine-100kn-vts-brand/ (accessed on 12 January 2023).
37. *ASTM E384-10*; Standard Test Method for Knoop and Vickers Hardness of Materials. ASTM International: West Conshohocken, PA, USA, 2010.
38. *ASTM G99-05*; Standard test method for wear testing with a pin-on-disk apparatus. ASTM International: West Conshohocken, PA, USA, 2010.
39. Bhansali, K.J.; Mehrabian, R. Abrasive wear of aluminum-matrix composites. *JOM* **1982**, *34*, 30–34. [CrossRef]

40. Kumar, N.M.; Kumaran, S.S.; Kumaraswamidhas, L. Aerospace application on Al 2618 with reinforced–Si3N4, AlN and ZrB2 in-situ composites. *J. Alloys Compd.* **2016**, *672*, 238–250. [CrossRef]
41. Santosh, R.N.; Sarojini, J.; Lakshmi, V. Enhancing the Mechanical Properties of Metal Matrix Composite by Reinforcing Aluminium 6063 with Sic & Graphite. *Int. J. Eng. Res. Technol. (IJERT)* **2018**, *6*, IJERTCONV6IS16014.
42. Hassan, M.K.; Redhwi, A.M.; Mohamed, A.F.; Backar, A.H.; Abdellah, M.Y. Investigation of Erosion/Corrosion Behavior of GRP under Harsh Operating Conditions. *Polymers* **2022**, *14*, 5388. [CrossRef] [PubMed]
43. Chen, H.S.; Wang, W.X.; Li, Y.L.; Zhang, P.; Nie, H.H.; Wu, Q.C. The design, microstructure and tensile properties of B4C particulate reinforced 6061Al neutron absorber composites. *J. Alloys Compd.* **2015**, *632*, 23–29. [CrossRef]
44. Kanth, U.R.; Rao, P.; Krishna, M. Mechanical behaviour of fly ash/SiC particles reinforced Al-Zn alloy-based metal matrix composites fabricated by stir casting method. *J. Mater. Res. Technol.* **2019**, *8*, 737–744. [CrossRef]
45. Seah, K.; Sharma, S.; Girish, B. Effect of artificial ageing on the hardness of cast ZA-27/graphite particulate composites. *Mater. Des.* **1995**, *16*, 337–341. [CrossRef]
46. Baradeswaran, A.; Perumal, A. Study on mechanical and wear properties of Al 7075/Al2O3/graphite hybrid composites. *Compos. Part B Eng.* **2014**, *56*, 464–471. [CrossRef]
47. Abdellah, M.Y.; Hassan, M.K.; AlMalki, A.A.; Mohamed, A.F.; Backar, A.H. Finite Element Modelling of Wear Behaviors of Composite Laminated Structure. *Lubricants* **2022**, *10*, 317. [CrossRef]
48. Sarina, B.A.; Kai, T.A.; Kvithyld, A.; Thorvald, E.N.; Tangstad, M. Wetting of pure aluminium on graphite, SiC and Al2O3 in aluminium filtration. *Trans. Nonferrous Met. Soc. China* **2012**, *22*, 1930–1938.
49. Allison, J.E.; Cole, G. Metal-matrix composites in the automotive industry: Opportunities and challenges. *JoM* **1993**, *45*, 19–24. [CrossRef]
50. Dhingra, A.K.; Fishman, S. Interfaces in Metal-Matrix Composites. In Proceedings of the Symposium, New Orleans, LA, USA, 4–6 March 1986; Metallurgical Society, Inc.: Warrendale, PA, USA, 1986.
51. Brown, W.; Eiss, N., Jr.; McAdams, H. Chemical Mechanisms Contributing to Wear of Single-Crystal Sapphire on Steel. *J. Am. Ceram. Soc.* **1964**, *47*, 157–162. [CrossRef]

metals

MDPI

*Review*

# Grain Refinement of Aluminum and Aluminum Alloys by Sc and Zr

**Zhiguo Lei, Shengping Wen *, Hui Huang, Wu Wei and Zuoren Nie**

Faculty of Materials and Manufacturing, Beijing University of Technology, Beijing 100124, China
* Correspondence: wensp@bjut.edu.cn

**Abstract:** Rare earth elements and transition elements are widely used in aluminum alloys, magnesium alloys and copper alloys due to their unique microalloying effects. With regard to in-depth research on the grain refinement characteristics of rare earth elements and transition elements, the combination of grain refinement and microalloying in the master alloys has a great impact on the theories and technical research of refinement, and the broadening of the application field of master alloys. This paper first summarizes the grain refinement mechanisms and analyzes the effects of rare earth elements and transition elements on the grain refinement of aluminum and aluminum alloys, and summarizes the elements suitable for the preparation of Al-M master alloys and their refinement mechanism.

**Keywords:** aluminum alloys; Sc; Zr; heterogeneous nucleation; edge to edge matching model; grain refinement; microalloying

## 1. Introduction

Aluminum and its alloys are highly attractive engineering materials for structural applications in many industries due to their low density, high specific strength and good corrosion resistance [1–3]. With the development of material technology, higher requirements have been put forward for the comprehensive properties of aluminum alloys, namely to obtain aluminum alloys with higher strength and better thermal stability [4]. The properties of aluminum alloys are controlled by a number of microstructural characteristics, such as grain size. Grain refinement, by promoting nucleation [5,6] and growth restriction [7], obtains a fine equiaxed crystal structure of α-Al, eliminates columnar structure, ensures uniform mechanical properties, improves formability and machinability, reduces casting defects and tearing tendency, and enhances subsequent machining response [8–10].

The grain refinement of commercial aluminum and aluminum alloys is mainly regulated by grain refiners [11]. Grain refiners are a type of master alloys. When the grain refining master alloys are added to melt, a large number of heterogeneous nucleating cores are released after melting. Those cores can affect the nucleating process of melt crystallization by acting as external cores during melt solidification, and play the role of refining grains. The most popular grain refiners are Al-Ti-B master alloys [12,13]. However, Al-Ti-B has no other effect after grain refinement, such as microalloying. Based on the Edge To Edge matching model [14,15] (E2E model), the coherent relationship between the two phases was determined by calculating the mismatch degree of the close-packed crystal plane and the close-packed crystal direction. A class of heterogeneous phase $Al_2RE$ (Rare Earth element) with a refining effect was screened for magnesium alloys. This phase can also improve the thermal stability of the microstructure [16–19]. In 2009, Qin et al. [16] added 0.6%~1.0 wt.%Al to Mg-10Y alloy, and the $Al_2Y$ formed by Y and Al in the melt reduced the grain size of the alloy from 180 μm to 36 μm. The size reduction rate was up to 80%, and the refinement effect was almost similar to that of adding Zr. $Al_2Y$ is a compound with high thermal stability, and the microstructure thermal stability of $Al_2Y$ is stronger than that of Zr-added magnesium alloy due to its uniform distribution within the grain and at the grain boundary, which can effectively hinder grain boundary migration.

Considering the heterogeneous phase $Al_2RE$, which has the function of refining magnesium alloy grains and microalloying, we draw on this idea and plan to design an Al-M (Metal) master alloy for the field of aluminum alloys, which both refines grain and improves thermal stability of the microstructure. Transition element Hf is a good example. When Hf is added to pure aluminum, the primary phase $Al_3Hf$ with $L1_2$ structure can be formed, and the primary phase can also pin the grain boundary after heterogeneous nucleation [20]. Sc, Zr and Er are often used as microalloying elements combined with Al to form the $Al_3M$ phase to improve the mechanical properties and heat resistance of aluminum alloys [21–26]. Literature review shows that rare earth elements and transition elements can form the $Al_3M$ phase with Al after solidification to refine the $\alpha$-Al grains of aluminum castings, improve the microstructure and obtain equiaxed grains [20,27–33]. Therefore, it is of great practical significance to study the grain refinement law, refinement mechanism and factors affecting the refinement effect of rare earth elements and transition elements to supplement the grain refinement theory and expand the application field of rare earth elements and transition elements. The application range of Al-M master alloys can be further expanded by combining the refined grains of elements with microalloying.

In this paper, the current mainstream grain refinement theory is summarized. Based on the generalized grain refinement theory [34–41], the refining mechanism and refining effect of rare earth elements and transition elements on the grains of aluminum and aluminum alloy are analyzed. The elements suitable for the preparation of an Al-M master alloy and their refining mechanism and influencing factors of refinement effect are summarized. Finally, the problems existing in the research process of Al-M intermediate alloys are presented.

## 2. The Mechanism of Grain Refinement

Based on a large number of studies [12,28,34–36,38,42–45], the theory of grain refinement on grain refinement of aluminum alloy refiners can be divided into constitutional supercooling of solute atoms and heterogeneous nucleation. Grain refiners refine grains mainly by promoting grain nucleation or inhibiting grain growth. Constitutional supercooling provides subcooling and inhibits grain growth. Heterogeneous phases provide a large number of nucleation sites, improve the nucleation rate and eventually lead to grain refinement. Both mechanisms are explained in detail below.

### *2.1. Constitutional Supercooling*

During the solidification process of alloys, the segregation of supersaturated solute atoms from the grains leads to a decrease in constitutional supercooling at the front of the solid–liquid interface and the inhibition of grain growth. This phenomenon is called compositional undercooling [36]. When the supercooling degree ($\Delta T_C$) provided by solute segregation is greater than the critical subcooling degree ($\Delta T_n$) required for nucleation, migration at the solid–liquid interface is hindered, and continued growth of original grains is also inhibited, which induces grain refinement. The ability of solute elements to inhibit grain growth is usually quantified by the Growth Restriction Factor (GRF for short, also known as $Q$). Easton and Stjohn [36] used GRF to quantitatively describe the inhibition effect of different solute elements on grain growth in cast alloys. The expression for calculating the growth restriction factor is as follows:

$$Q = mC(k-1) \tag{1}$$

where $m$ is the liquidus slope of the aluminum-rich end of the binary alloy, $C$ is the solubility of the solute and $k$ is the solute partition coefficient (e.g., $k = \frac{C_S}{C_L}$, $S$ represents the $\alpha$-Al phase and $L$ represents the liquid phase, $C_S$ is the maximum solid solubility of the aluminum-rich end of the binary alloy and $C_L$ is represents peritectic point or eutectic point).

Generally speaking, for the liquidus slope of the Al-rich end of eutectic reactive binary alloys, $m$ is negative, and the $k$ value is less than 1. For peritectic reaction type binary alloy, $m$ is positive, and the $k$ value is greater than 1. Plug $m$, $k$ and $C$ into Formula 1, so that $Q$ is always positive. For solute elements that do not interact with each other, their influence on the solidification process can be measured by the sum of the growth limiting factors of various solutes ($\sum Q$). Table 1 shows the theoretical growth-limiting factors of different solute elements during alloy solidification. According to the Table 1, the Ti element has the highest value of $Q/C$, and has the best grain growth inhibition effect and can produce the best grain refinement effect under the same mass percentage content. Johnsson [46] studied the subcooling of Ti components and found that the larger $Q$, the better the grain refinement effect. When $Q$ is 7, the grain size of pure aluminum is 250 μm. When $Q$ increases to 23, the grain size decreases to 120 μm.

**Table 1.** Parameters underlying the calculation of $Q$ for different solutes in aluminum [47].

| Element | System | $m$ (K/wt.%) | $K$ | $C$ (wt.%) | $Q/C$ (K/wt.%) |
|---|---|---|---|---|---|
| Cu | eutectic | −3.4 | 0.17 | 33.2 | 2.8 |
| Ni | eutectic | −3.3 | 0.007 | 6 | 3.3 |
| Mg | eutectic | −6.2 | 0.51 | 3.4 | 3.0 |
| Mn | eutectic | −1.6 | 0.94 | 1.9 | 0.1 |
| Si | eutectic | −6.6 | 0.11 | 12.6 | 5.9 |
| Fe | eutectic | −3.0 | 0.02 | 1.8 | 2.9 |
| Er | eutectic | −0.91 | 0.03 | 6 | 0.88 |
| Sc | eutectic | −9.4 | 0.59 | 0.58 | 3.90 |
| Ti | peritectic | 7.8 | 33.3 | 0.15 | 220 |
| Cr | peritectic | 3.5 | 2.0 | 0.4 | 3.5 |
| Ta | peritectic | 70.0 | 7.8 | 0.10 | 105 |
| V | peritectic | 10.0 | 4.0 | 0.1 | 30.0 |
| Hf | peritectic | 8.0 | 2.4 | 0.5 | 11.2 |
| Mo | peritectic | 5.0 | 2.5 | 0.1 | 7.5 |
| Nb | peritectic | 13.3 | 1.5 | 0.15 | 6.6 |
| Zr | peritectic | 4.5 | 2.5 | 0.11 | 6.8 |

In addition to the data found in reference [47], this document strengthens the table with more information.

### 2.2. *Heterogeneous Nucleation*

In addition to the constitutional supercooling mechanism of solute atoms, elements also exist in the form of phases, which should also be considered as heterogeneous nucleation. The mechanism of heterogeneous nucleation is based on the crystallographic relationship between the heterogeneous phase and matrix. There are three prevailing views: the lattice matching principle [40], two-dimensional matching model [39] and Edge To Edge matching model [14,15,43,48].

#### 2.2.1. Lattice Matching Principle

Traditional heterogeneous nucleation theory holds that the distinguishing condition for heterogeneous particles to exert nucleation effect is that there is a small interfacial energy between the particle and matrix phase. The mismatch degree between the heterogeneous phase and matrix, the electrostatic potential, and the morphology, size and size distribution of the heterogeneous phase are all factors affecting the interfacial energy, among which the mismatch degree is the dominant factor affecting the nucleation efficiency of the heterogeneous phase. The mismatch degree between the heterogeneous phase and the matrix is widely used to measure its refining ability or to find new nucleating particles. According to the solidification theory, the smaller the lattice mismatch between particles and matrix, the lower the interfacial energy of two phases and the stronger the heterogeneous nucleation effect of particles. Initially, based on lattice constants, Turnbull and Vonnegut [40] theoretically studied the lattice mismatch degree between matrix and nucleated particle, and proposed the formula of mismatch degree ($\delta$):

$$\delta = \frac{\Delta a_0}{a_0} \tag{2}$$

where $\Delta a_0$ is the difference between the lattice constant of the matrix and the particle on the low index crystal plane and $a_0$ is the lattice constant of the matrix.

The model points out that when $\delta \leq 15\%$, the nucleation particle can produce a refining effect, otherwise there is no refining ability.

Based on the lattice matching theory, Bramfitt [39] established two-dimensional lattice matching model to make the low index crystal planes of nucleating particles and matrix phases coincide, which can be applied to the calculation of the mismatch between two phases of crystal planes with different atomic arrangement. The formula for calculating the two-dimensional mismatch ($\delta\frac{(hkl)_s}{(hkl)_n}$) is as follows:

$$\delta\frac{(hkl)_s}{(hkl)_n} = \frac{1}{3}\sum_{i=1}^{3}\frac{\left|d[uvw]_s^i \cos\theta - d[uvw]_n^i\right|}{d[uvw]_n^i} \times 100\% \quad (3)$$

where $s$ is the matrix and $n$ is the nucleating particle. $(hkl)$ is low index crystal facet; $[uvw]$ is the low index orientation on the low index crystal plane; $d[uvw]_s$ and $d[uvw]_n$ are the atomic spacing along the corresponding crystal plane and the crystal direction, respectively; $\theta$ is the angle between the crystal direction of $[uvw]_s$ and $[uvw]_n$. The theory states that heterogeneous nucleation is effective when $\delta < 6\%$. If $\delta = 6\%–12\%$, it has medium efficacy; if it is more than 12%, granules cannot be used as heterogeneous nucleation cores. The face mismatch degree between $Al_4C_3$ and Mg matrix is 4.05% [46]. Subsequent experiments show that $Al_4C_3$ particles can significantly refine the as-cast grains of magnesium alloy, which fully verifies the effectiveness of the surface matching model.

Based on previous research results, Zhang et al. [14,15] proposed the E2E model to consider the mismatch degree from the perspective of atomic row (crystal direction) at the phase interface. The key assumption of the E2E model is that the nucleation substrate always occurs at the orientation with the least mismatch when the atomic row at the two-phase interface is parallel and the mismatch degree is minimum. For the requirement of atomic columns, the atomic columns satisfying the hypothesis should first be densely or nearly densely packed; secondly, the atomic columns can be arranged in straight lines (abbreviated as S) or zig-zag lines (abbreviated as Z), as shown in Figure 1. Finally, the lines of straight atoms correspond to the lines of straight atoms, and the lines of sawtooth atoms correspond to the lines of sawtooth atoms.

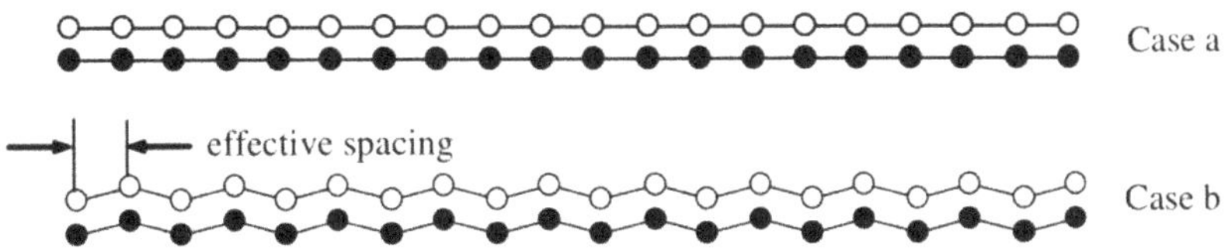


**Figure 1.** Schematic illustration of two different types of atom row matching at the interface. Reprinted with permission from [15], 2005, Elsevier.

The E2E model quantifies the heterogeneous nucleation ability of particles using two parameters, namely interatomic spacing mismatch (it also known as $f_r$) and interplanar spacing mismatch (it also known as $f_d$), which correspond to the orientation and plane of crystal structure, respectively. When the $f_d$ and $f_r$ of nucleation substrate and primary crystal nucleus in a certain orientation are less than 6% and 10%, respectively, the nucleation substrate is an effective heterogeneous core. The atomic spacing mismatch ($f_r$) and crystal plane spacing mismatch ($f_d$) between phases can be calculated using the formula. The specific calculation formula is as follows:

$$f_d = \left|\frac{d_m - d_p}{d_m}\right| \times 100\% \quad (4)$$

$$f_r = \left| \frac{r_m - r_p}{r_m} \right| \times 100\% \quad (5)$$

where $d_m$ is the crystal plane spacing of matrix phase densely packed or nearly densely packed, $d_p$ is the crystal plane spacing of heterogeneous phase densely packed or nearly densely packed, $r_m$ is the atomic spacing of the matrix phase packed or nearly packed and $r_p$ is the atomic spacing of the heterogeneous phase packed or nearly packed.

The reliability of E2E model to predict refinement potential has been verified [42,43,48]. Wang [28] calculated the possible orientation relationship between $Al_3Zr$ phase and α-Al using the E2E model, and found that there are 6 groups of orientation relationships between $Al_3Zr$ and α-Al, and the mismatch degree is less than 6%. The orientation relationships of $[1\bar{1}0]_{Al_3Zr}(114)_{Al_3Zr}//[101]_{Al}(11\bar{1})_{Al}$ and $[40\bar{1}]_{Al_3Zr}(114)_{Al_3Zr}//[1\bar{1}0]_{Al}(11\bar{1})_{Al}$ were verified using EBSD (Electron Backscatter Diffraction), indicating that $Al_3Zr$ is indeed a heterogeneous core promoting α-Al nucleation. Li et al. [43] used the E2E model to predict whether $LaB_6$ had the potential to be a heterogeneous nucleating substrate of nascent $Al_{13}Mn_4Si_8$, and reached similar conclusions.

### 2.2.2. Free Growth Model

In addition to the mismatching degree of interfacial atomic row, heterogeneous particle size is also a key factor affecting the heterogeneous nucleation efficiency [37,49–51]. When particle size met the conditions, the undercooling degree could be reduced to provide driving force and promote heterogeneous nucleation.

$$d_n = \frac{4\sigma_{SL}}{\Delta S_v \Delta T_n} \quad (6)$$

where $d_n$ is particle diameter (assumed to be spherical); $\sigma_{SL}$ is solid–liquid interface energy, about 158 $mJ \cdot m^{-2}$ [37]; $\Delta S_v$ is the melting entropy per unit volume ($\Delta S_v$ is the ratio of melting potential heat $L_v$ to melting point $T_m$), which is determined to be $1.112 \times 10^6$ $J \cdot K^{-1} \cdot m^{-3}$ [37] in the experiment; $\Delta T_n$ is the critical degree of nucleation subcooling. Greer et al. [37,52,53] used the free growth model to study the grain refining effect of Al-Ti-B intermediate alloy, and found for the first time the quantitative relationship between the size of $TiB_2$ particles and the final grain refining effect. Greer et al. believed that larger $TiB_2$ particles preferentially act as nucleation cores to promote α-Al nucleation in the crystallization process. With the decrease in the melt temperature, the degree of supercooling increases, and smaller $TiB_2$ particles begin to produce a heterogeneous nucleation effect. At the same time, the latent heat release of crystalline grains leads to a slow decrease in temperature. No new grains are formed, and the smaller $TiB_2$ particles cannot serve as an effective nucleating core. Similarly, Yu et al. [54] analyzed the size change of $Al_3Ti$ in ECAP (Equal Channel Angular Pressing). After 10 ECAP deformations, when the size of $Al_3Ti$ particles remained in the optimal size range of 18–22 μm, the grain refinement performance of pure Al in the study object was the best and stable.

## 3. Al-M Master Alloy

Rare earth elements represented by Sc and Er, and transition elements represented by Zr are added to aluminum alloy in the form of master alloys. Nano-scale $Al_3M$ phase precipitated from the supersaturated solid solution of α-Al plays an important role in improving the recrystallization temperature of the aluminum alloy, and its mechanical properties and heat resistance [55]. A large number of studies [28,56–60] have shown that the three elements Sc, Zr and Er can also improve the casting microstructure and refine the grain. Considering that refiners in the field of magnesium alloy can play the role of refining grain and microalloying, we reference this idea. The existing research results of grain refinement of rare earth elements and transition elements are analyzed, and the elements suitable for developing Al-M master alloys with microalloying and grain refinement are summarized. Finally, the existing problems of Al-M master alloys are described.

### 3.1. Al-Sc

The Sc element is often used as a microalloying element to improve the microstructure and properties of aluminum alloys. When Sc is added to aluminum alloy, after solid solution and aging treatment, the supersaturated solid solution α-Al is decomposed into $Al_3Sc$ precipitated phase with $L1_2$ structure. This is evenly and finely dispersed in the aluminum matrix, increasing the resistance of dislocation slip and climbing, impeding the migration of grain boundaries and improving the effect of recrystallization temperature and microstructure thermal stability [21,22,33,55,61,62]. In addition, Sc has the effect of refining α-Al grains [30–33,57,63–65].

Sc exists in aluminum alloys in two forms, solute atoms and $Al_3Sc$ particles [21,57,59], and the mechanism of refinement is also different. With regard to solute atoms, during the crystallization process, supersaturated Sc atoms are drained from the grains and biased at the front end of the solid–liquid interface, resulting in supercooling of the components, providing a degree of supercooling and inhibiting the growth of grains [57,59]. This refinement effect is due to the fact that the growth limiting factor of Sc is $3.9C_0$. As the core of heterogeneous nucleation [21,32], the $Al_3Sc$ phase significantly improves the nucleation rate, promotes α-Al grain nucleation and leads to grain refinement. This is because the $Al_3Sc$ phase has an $L1_2$ structure with a lattice constant of 0.4104 nm [55], which is very similar to Al in terms of face-centered cubic structure and lattice constant (0.4049 nm), and there is a small lattice mismatch(1.32% [55]) between the two phases.

The grain refining mechanism of Sc is different, which leads to the difference in the refining effect of α-Al grains. According to the results of literature studies [57,65], the refining effect of $Al_3Sc$ is stronger than that of solute atoms. As can be seen from Figure 2a, the grain size of α-Al without addition of Sc element is 860 μm. When 0.2 wt.%Sc is added, the grain size of α-Al is 830 μm, the grain size is coarse and uneven (Figure 3a) and there is no significant reduction in grain size. The grain refining effect is gradually enhanced with the increase in Sc content within eutectic composition (Sc = 0.58 wt.% Figure 2b). The grain size of 0.5 wt.%Sc alloy is 700 μm, and the grain size reduction is 160 μm. The grain size reduction rate is 18.6%. When Sc content is increased beyond the eutectic point, grain size decreases more significantly. An addition of 0.7 wt.% Sc can refine the grain to 40 μm. Figure 3b shows that the grain size is fine and uniform, and the reduction rate reaches 95.3%. This difference in refinement results from refinement mechanism. According to the Al-Sc binary phase diagram (Figure 2b), when the addition amount of Sc is lower than that of eutectic component, Sc element exists in the matrix in the form of solute atoms, and component subcooling is used as the mechanism of grain refinement. Beyond the eutectic point, $Al_3Sc$ begins to be the main form of Sc. At the same time, the refinement mechanism changes from the subcooling of solute atoms to heterogeneous nucleation. Heterogeneous nucleation has a stronger refinement effect and can significantly refine α-Al grains.

However, the refinement theory and optimization of refinement effect of $Al_3Sc$ need further study. The current understanding of grain refinement mechanism still remains the lattice mismatch proposed in 1952. In 2005, Zhang [66–68] proposed the E2E model to deeply understand the crystallographic relationship between heterogeneous phases and matrix. In 2021, Yan [65] found that there was a good orientation relationship between $Al_3Sc$ and α-Al:

$$[011]_{Al}//[011]_{Al_3Sc}\&(\bar{2}00)_{Al}//(\bar{2}00)_{Al_3Sc} \text{ and } [\bar{1}12]_{Al}//[\bar{1}12]_{Al_3Sc}\&(\bar{1}1\bar{1})_{Al}//(\bar{1}1\bar{1})_{Al_3Sc}$$

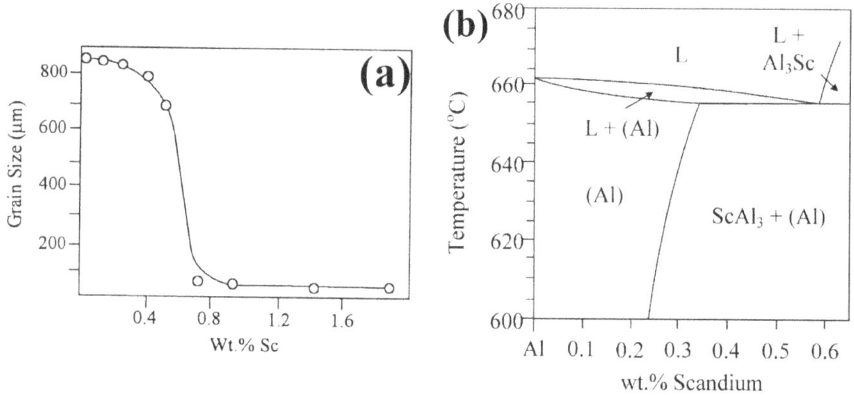


**Figure 2.** Curve of grain size addition of Sc [57] (**a**) and phase diagram of Al-Sc [57] (**b**). Reprinted with permission from [57], 1998, Elsevier.

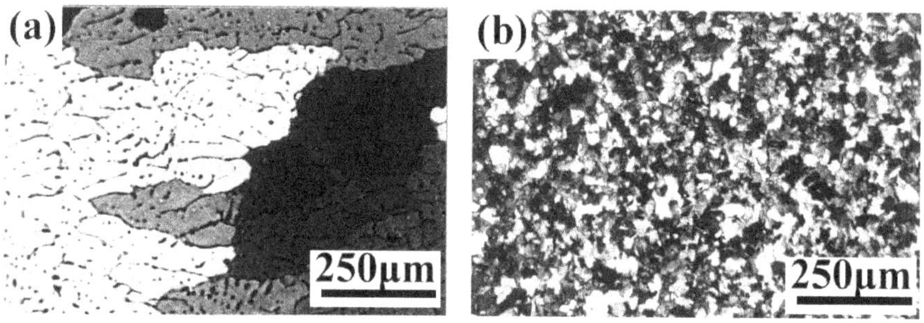


**Figure 3.** Optical microstructure (OM) structure [57]: (**a**) Al-0.2Sc, (**b**) Al-0.7Sc. Reprinted with permission from [57], 1998, Elsevier.

However, the crystallographic study of $Al_3Sc$ grain refinement has not been studied enough. In addition, the grain refinement effect is affected by the morphology, and the primary phase $Al_3Sc$ has many morphologies [69,70]. Sun [69] studied the relationship between melting conditions and the morphology of primary phase $Al_3Sc$, and found that cooling rate was the key factor affecting the morphology evolution of $Al_3Sc$. Melting temperature had a greater effect on the dendrite degree, and holding time had a greater effect on the primary phase size. Moderate melting temperature and rapid cooling result in the diversity of primary $Al_3Sc$ phase morphology, and this $Al_3Sc$ has the best refining effect on the studied alloys. Higher melting temperature and slower cooling rate will result in the primary $Al_3Sc$ phase with butterfly shape and cracked cube with pointed arrows at the corners, and the refining effect of this master alloy will be significantly reduced. Very low cooling rate and high melting temperature lead to dendrite formation of primary $Al_3Sc$ phase with the worst thinning effect and microstructure inheritance.

### *3.2. Al-Zr*

Zr, as a transition group element, is added to magnesium alloy in the form of Mg-Zr master alloy, which can produce strong grain refinement effect [71,72]. Zr has also been reported to refine α-Al grains [28–30,63,64].

Zr exists in different forms in aluminum alloys [28], and its refining mechanism is also different. According to the Al-Zr phase diagram [73], the solid solubility of Zr is 0.11 wt.%, and Zr exists in the form of solute atoms at this concentration. According to Table 1, the GRF of Zr is $6.8C_0$, and the segregation at the solid–liquid interface during crystallization leads to the undercooling of the components. This reduces the undercooling degree, inhibiting the grain growth and refining the grains. In the presence of the $Al_3Zr$ phase, the mechanism of grain refinement is heterogeneous nucleation, which can be explained by Zhang Mingxing's Edge To Edge model [41,67,68]. The $Al_3Zr$ phase formed during the solidification process is a $D0_{23}$ crystal structure, belonging to the tetragonal crystal structure, with lattice constants of a = 0.4007 nm and c = 1.7286 nm. Each cell contains 12 Al atoms and 4 Zr atoms [74].

The most densely packed facet of the $Al_3Zr$ phase is $\{114\}_{Al_3Zr}$ that contains three densely packed atomic columns($< 22\bar{1} >^{S}_{Al_3Zr}$, $< 110 >^{Z}_{Al_3Zr}$ and $< 40\bar{1} >^{Z}_{Al_3Zr}$). For convenience, the superscript "Straight (for short, S)" and "Zig-Zag(for short, Z)" are used to distinguish the straight row atomic arrangement from the sawtooth row atomic arrangement, as shown in Figure 4a,b. A sub-densely packed plane is $\{020\}_{Al_3Zr}$, which contains only one dense row of atoms ($< 40\bar{1} >^{Z}_{Al_3Zr}$). The third densely packed surface of the $Al_3Zr$ phase is $\{220\}_{Al_3Zr}$, which contains two densely packed atomic columns ($< 22\bar{1} >^{S}_{Al_3Zr}$ and $< 110 >^{Z}_{Al_3Zr}$).

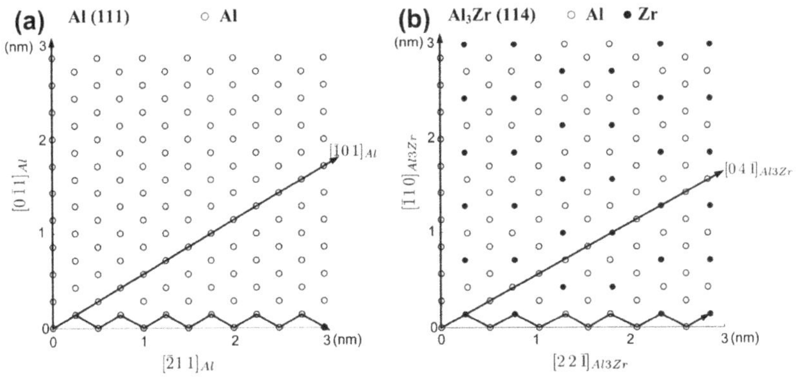


**Figure 4.** Schematic diagram of the arrangement of atoms [28]: (**a**) Al, (**b**) $Al_3Zr$. Reprinted with permission from [28], 2013, Elsevier.

Aluminum is a face-centered cubic structure with a lattice constant of a = 0.4049 nm and 4 Al atoms in the cell [74]. The most densely packed surface of Al is $\{111\}_{Al}$. It has two closely packed columns of atoms: $< 110 >^{Z}_{Al}$ and $< \bar{2}11 >^{S}_{Al}$. Furthermore, $\{020\}_{Al}$ is a sub-dense plane with only one dense row of atoms ($< 110 >^{Z}_{Al}$). The third closest plane is $\{220\}_{Al}$ containing $< 110 >^{Z}_{Al}$ and $< \bar{2}11 >^{S}_{Al}$.

The interatomic spacing mismatch and interplanar spacing mismatch between $Al_3Zr$ and α-Al were calculated respectively, and the results are shown in Table 2. Table 3 lists the orientation relationships that meet the conditions of interatomic spacing mismatch less than 10% and interplanar spacing mismatch less than 6%. EBSD results prove the existence of two groups of orientation relationships (Figure 5):

$$[101]_{Al}(11\bar{1})_{Al}//[1\bar{1}0]_{Al_3Zr}(114)_{Al_3Zr} \text{ and } [1\bar{1}0]_{Al}(11\bar{1})_{Al}//[40\bar{1}]_{Al_3Zr}(114)_{Al_3Zr}$$

**Table 2.** Result of calculations [28].

| Matching Crystal Direction | $f_r$ | Matching Crystal Plane | $f_d$ |
|---|---|---|---|
| | | $\{111\}_{Al}//\{114\}_{Al_3Zr}$ | 1% |
| $<110>^S_{Al}$ // $<110>^S_{Al_3Zr}$ | 1% | $\{111\}_{Al}//\{020\}_{Al_3Zr}$ | 17% |
| | | $\{111\}_{Al}//\{220\}_{Al_3Zr}$ | 65% |
| | | $\{020\}_{Al}//\{114\}_{Al_3Zr}$ | 15% |
| $<110>^S_{Al}$ // $<40\bar{1}>^S_{Al_3Zr}$ | 2% | $\{020\}_{Al}//\{020\}_{Al_3Zr}$ | 1% |
| | | $\{020\}_{Al}//\{220\}_{Al_3Zr}$ | 42% |
| | | $\{022\}_{Al}//\{114\}_{Al_3Zr}$ | 40% |
| $<\bar{2}11>^Z_{Al}$ // $<22\bar{1}>^Z_{Al_3Zr}$ | 4% | $\{022\}_{Al}//\{020\}_{Al_3Zr}$ | 29% |
| | | $\{020\}_{Al}//\{220\}_{Al_3Zr}$ | 1% |

**Table 3.** Orientation relationship between $Al_3Zr$ phase and α-Al [28].

| Matching Crystal Direction | $f_r$ | Matching Crystal Plane | $f_d$ |
|---|---|---|---|
| $<110>^S_{Al}$ // $<110>^S_{Al_3Zr}$ | 1% | $\{111\}_{Al}//\{114\}_{Al_3Zr}$ | 1% |
| | | $\{022\}_{Al}//\{220\}_{Al_3Zr}$ | 1% |
| $<110>^S_{Al}$ // $<40\bar{1}>^S_{Al_3Zr}$ | 2% | $\{111\}_{Al}//\{114\}_{Al_3Zr}$ | 1% |
| | | $\{020\}_{Al}//\{020\}_{Al_3Zr}$ | 1% |
| $<\bar{2}11>^Z_{Al}$ // $<22\bar{1}>^Z_{Al_3Zr}$ | 4% | $\{111\}_{Al}//\{114\}_{Al_3Zr}$ | 1% |
| | | $\{022\}_{Al}//\{220\}_{Al_3Zr}$ | 1% |

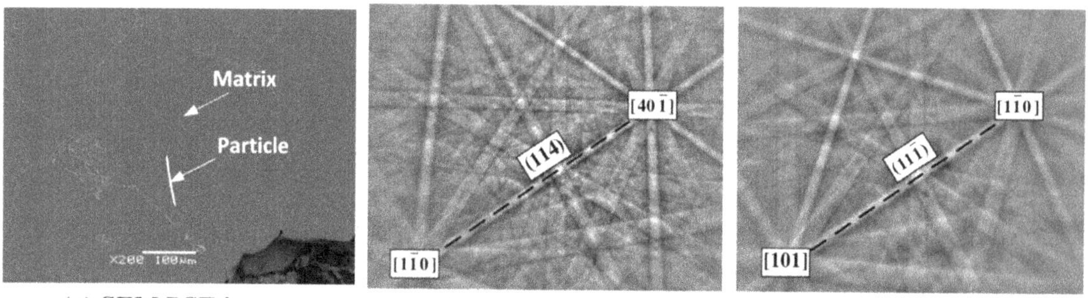


**(a) SEM BSE image** **(b) $Al_3Zr$ Pattern** **(c) Al Pattern**

**Figure 5.** Backscattered electronic (BSE) image of the microstructure and EBSD Kikuchi pattern of $Al_3Zr$ and Al [28]. Reprinted with permission from [28], 2013, Elsevier.

The grain refinement mechanisms of solute atom Zr and $Al_3Zr$ are different, which leads to the difference in grain refinement effect. The refining effect of $Al_3Zr$ phase is much better than that of solute atomic Zr, which occupies a dominant position. Wang et al. [28] studied the effect of Zr addition on grain refinement of pure aluminum. The results show that the equilibrium solidification microstructure is obtained by controlling the cooling rate at 1 K/s when the addition amount is 0.1 wt.% (Figure 6b,e). It was found that the pure Al grain size is reduced from 1100 μm to 1050 μm (Figure 6a,e), and the grain size reduction rate is 4.5%. The grain size of pure aluminum is 500 μm and the size reduction rate is 54.5% when the addition amount is increased to 0.2 wt.% (Figure 6c,e). The addition was further increased to 0.5 wt.%, and the grain size was reduced to 350 μm (Figure 6e). This difference in refinement results from the change of refinement mechanism. When 0.1 wt.%Zr is added, the constitutional undercooling of solute atomic Zr acts as a refinement mechanism. If the addition amount of Zr is above 0.1 wt.%, $Al_3Zr$ is detected by XRD and serves as the core of α-Al nucleation.

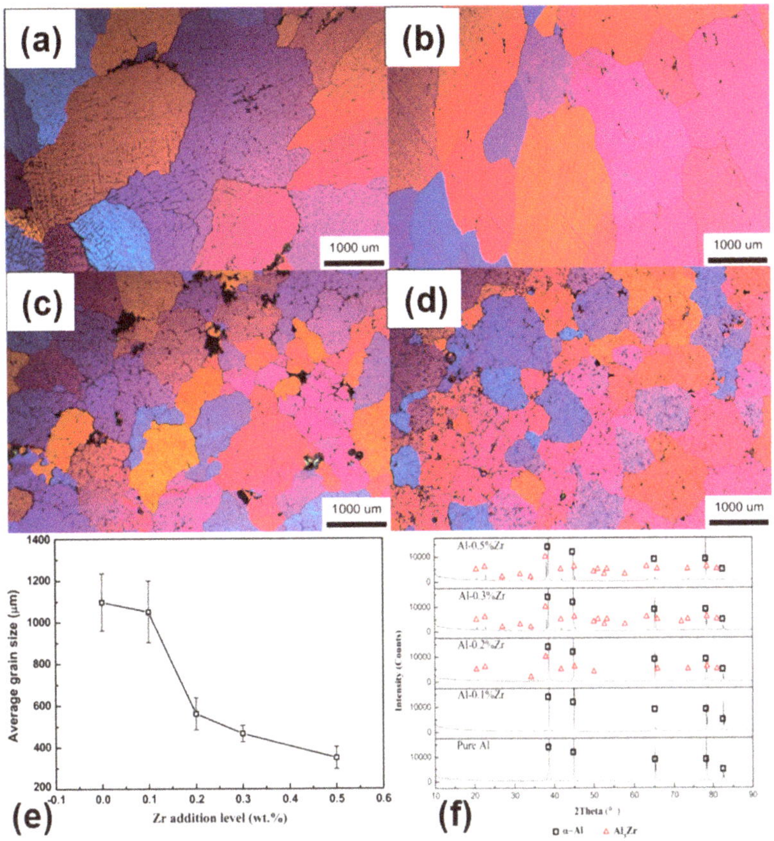

Figure 6. Results of grain refinement [28]: (a) 0% Zr, (b) 0.1% Zr, (c) 0.2% Zr, (d) 0.3% Zr, (e) relationship of Al grain size and Zr addition level, (f) XRD result of as-cast Al. Reprinted with permission from [28], 2013, Elsevier.

The refining effect of heterogeneous phase is affected by its morphology [75,76]. Through literature research, it was found that the morphology of $Al_3Zr$ will change according to different casting temperature [76]. When the temperatures were 1123 K, 1173 K, 1223 K and 1273 K, the morphologies of the $Al_3Zr$ of spherical shape, tetragon shape, rod shape and fiber shape were observed, respectively. At present, the effect of different $Al_3Zr$ morphologies on the refinement results has not been reported. The study on the effect of morphology on refinement has a positive effect on the optimization of refinement ability.

### 3.3. Al-Er

The price of Er is only $\frac{1}{80}\sim\frac{1}{100}$ that of Sc [77], and the Er element is often added to aluminum alloy as an alternative element of Sc to produce microalloying and improve the microstructure and mechanical properties of the alloy [26,27,78–80]. As a rare earth element, a refined grain is one of the characteristics of Er [27,60,81].

Er exists in aluminum alloy in two forms: solute atom and $Al_3Er$ phase. Based on the literature review [27,60,81], the solute atom Er can only refine the α-Al dendrite arms, while the primary $Al_3Er$ phase can refine the grains. Wen et al. [27] studied the effect of refining as-cast Al-Mg-Mn-Zr alloys with different additions of Er, and observed that the addition of 0.2 wt.%Er did not lead to grain refinement compared with the alloys without Er. However, as shown in Figure 7b, there does appear to be refinement in the dendritic substructure within the grains. The grain size of α-Al decreased significantly after adding 0.4 wt.%Er. By comparison with Figure 7a,c, the grain size of the alloy containing 0.4 wt.%Er is only half of that without the addition of the alloy. Increasing the addition of Er will only lead to slight decrease in grain size. Figure 8 shows the XRD diffraction patterns of the as-cast alloy with different Er content and the homogenized alloy with 0.4 wt.%Er added. The peak of $Al_3Er$ phase appeared only after 0.4 wt.%Er was added, and the diffraction peak still existed after homogenization. However, this indicates that 0.2 wt.% Er can be dissolved in the matrix. In summary, according to the Al-Er phase diagram [82], the maximum solid solubility of Er is 0.05 wt.%, and XRD results show that there is no formation of $Al_3Er$ in the alloy added 0.2 wt.%Er. In the solidification process, the supersaturated Er is segregated at the grain boundary, but the GRF of Er is $0.88C_0$, which is much smaller than that of Zr ($6.8C_0$) and Sc ($3.9C_0$), so it is difficult to produce a refinement effect. At 0.4 wt.%, the presence form of Er changes, with solute atoms and intermetallic compounds co-existing. Except for a small part of Er that is dissolved in the matrix, most Er is segregated at the front end of the solid–liquid interface. The aggregation of Er at the interface will make the composition fluctuation near the interface more drastic, because in the Al-Er binary phase diagram, the eutectic reaction temperature at the Al-rich end is 655 °C, and the eutectic composition is 6 wt.%. When the melt near the solid–liquid interface is cooled to 655 °C, $Al_3Er$ compounds can be formed in the melt with high erbium content (more than eutectic point composition) near the interface. These $Al_3Er$ compounds are directly extracted from the melt. When the primary $Al_3Er$ phase appears in the melt, there is still much liquid that has failed to crystallize and nucleate. The lattice type ($L1_2$) and lattice constant (a = 0.4215 nm) of $Al_3Er$ are very close to that of aluminum matrix (face-centered cubic structure, lattice constant a = 0.4049 nm), and the lattice mismatch is 4.10% [81]. The two phases are congruent and semi-congruent. Moreover, the melting point of $Al_3Er$ is 1067 °C, which is much higher than the melting point of the matrix (660 °C), and it has good stability after formation and will not melt back into the matrix. These characteristics ensure that the $Al_3Er$ phase has a good heterogeneous nucleation, so the primary $Al_3Er$ phase can serve as the nucleation core and play a role in refining grains when the melt continues to crystallize and nucleate.

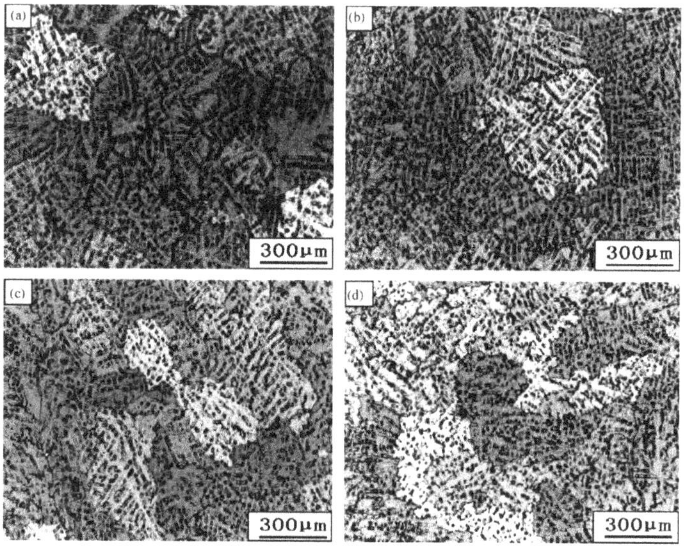


**Figure 7.** OM microstructure of as-cast Al-Mg-Mn-Zr-Er alloy [27]: (**a**) 0Er, (**b**) 0.2 wt.%Er, (**c**) 0.4 wt.%Er, (**d**) 0.6 wt.%Er. Reprinted with permission from [27], 2009, Elsevier.

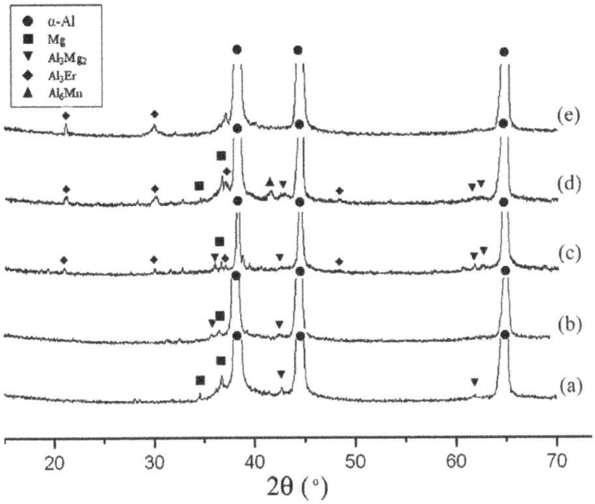


**Figure 8.** XRD diffraction patterns of as-cast alloys [27]: (**a**) 0Er, (**b**) 0.2 wt.%Er, (**c**) 0.4 wt.%Er, (**d**) 0.6 wt.%Er, (**e**) homogenized with 0.4 wt.%Er. Reprinted with permission from [27], 2009, Elsevier.

The primary phase has excellent thermal stability. Wen et al. [27] found that 0.2 wt.%Er was dissolved in the Al matrix as a solid solution, while other excess Er atoms were segregated at grain boundaries during solidification to form the main $Al_3Er$ phase. After homogenizing the cast alloy at 470 °C for 24 h (Figure 8), the morphology and structure of the primary $Al_3Er$ did not change, indicating that $Al_3Er$ could not dissolve into the Al matrix at this temperature because the melting point of $Al_3Er$ is much higher than the melting point of Al alloy. Although $Al_3Er$ can refine grains and has good thermal stability, Sc is an excellent candidate for intermediate alloys considering the crystal constant, lattice mismatch and growth limiting factor of $Al_3Sc$.

### 3.4. Al-Sc-Zr

Based on the previous analysis of refinement mechanism and refinement effect of elements, Sc, Er and Zr all have the effect of refining grains. By forming primary $Al_3M$ phase and using heterogeneous nucleation as grain refining mechanism, they can all be used as candidate elements for preparing Al-M intermediate alloys.

The compound addition of elements in pairs is a means to improve the refining effect and reduce the amount and cost of rare earth addition. Studies have shown that the addition of Sc and Zr can significantly refine α-Al grains in aluminum alloys [33,63]. Yin et al. [63] studied the effect of compound addition of Sc and Zr on the casting microstructure of Al-Mg alloy. The results show that, compared with adding Sc and Zr separately, adding 0.2 wt.%Sc and 0.1 wt.%Zr to Al-5Mg alloy at the same time, the grain size of the alloy decreases more obviously, and the equiaxed grains are more uniform. Zou et al. [83] also found through experiments that the compound addition of trace Sc and Zr can better refine the grain of cast aluminum alloy. In addition, the combination of Sc and Zr elements can reduce the use of Sc elements, thus reducing the production cost. Meng et al. [64] studied the effect of rare earth Sc and Zr on the microstructure of Al-Mg-Si-Cu-Cr-V alloy, and found that the improvement effect of 0.13 wt.%Zr and 0.03 wt.%Sc added at the same time is almost the same as that of 0.15 wt.%Sc. Esmaeili et al. [84] found similar grain refinement in Al-Mg-Si-Sc-Zr alloys and reached the same conclusion. It is generally believed that the compound addition of Sc and Zr can produce a synergistic effect, that is, the precipitation strengthening effect of the $Al_3(Sc, Zr)$ phase formed by the decomposition of supersaturated solid solution is introduced to strengthen the matrix better [85–87]. Recent studies [88] indicate that controlling the mass ratio of

Sc and Zr can change the primary phase type. Wang [88] found $Al_3(Sc, Zr)$ in the cast aluminum alloy after refining, where the mass ratio of Sc and Zr is 1, and the addition amount is 0.2 wt.%. When the mass ratio of Sc to Zr was adjusted to 4 and the addition amount of Zr was 0.1 wt.%, the cast microstructure was only $Al_3Sc$, and their composite phase was not detected [89]. The mass ratio of Sc to Zr was adjusted to 0.5, the addition amount of Zr was 0.2 wt.%, the cast microstructure was only $Al_3Sc$, and their composite phase was not detected [90]. Compared to $Al_3Sc$, $Al_3(Sc, Zr)$ is also an $L1_2$ structure with a smaller lattice mismatch (1.15%) with Al [91].Further analysis shows that by regulating the mass ratio of Sc and Zr, the primary phase type changes and the mechanism of grain refinement is different, which appears in Al-V-B [92].Whether the size, morphology and distribution will change needs to be further studied.

## 4. Prospects

Aluminum alloy with light weight, excellent mechanical properties, corrosion resistance and conductive properties are used in aerospace, ships, transportation and other fields. Due to the influence of casting process parameters and environmental factors, the organization and composition of cast aluminum parts are not uniform, which limits the expansion of the application range of aluminum alloy. In order to solve this problem, based on the idea of screening out $Al_2RE$ in magnesium alloy, which can refine grain and improve thermal stability of microstructure, this paper proposes the idea of preparing a new Al-M master alloys with grain refinement and a microalloying effect, which can not only solve the uneven structure and composition of casting, but also produce a microalloying effect after refining. This further improves the heat resistance of aluminum alloy. The selection of elements suitable for the preparation of new master alloys, the discussion of the refinement mechanism of elements, the analysis of the refinement effect of elements and the summary of factors affecting the refinement effect have very important theoretical significance for understanding the nature of the chemical method affecting the refinement of the grain, and its application in industrial production and guidance of the development of intermediate alloys, but also has practical application value. We selected Sc and Zr elements as reference objects for Al-M master alloys, but there are still many problems:

(1) The refinement mechanism of $Al_3Sc$ needs further analysis. At present, the understanding of the refinement mechanism is still in the crystal structure, lattice constant and lattice mismatch. With the help of E2E model, it is helpful to explore the crystallographic relationship between $Al_3Sc$ and $\alpha$-Al and to improve the refinement theory. Sc is a eutectic reaction type element, and the primary $Al_3Sc$ phase in the eutectic structure during the solidification process is isolated at the grain boundary. The influence of the size, distribution and number density of the primary phase on the refining effect is still unclear.

(2) The primary $Al_3Zr$ phase will have different morphologies when the melting conditions are different. The law and mechanism of the influence of morphologies on the refining effect need to be explored using experiments. The size and distribution of the primary phase are also factors to be considered.

(3) The analysis of the strength of the grain refinement effect of Sc and Zr elements can provide data in support of the optimization of the refinement effect and the design of ternary alloy composition.

(4) When the pairwise addition of Sc and Zr elements to prepare ternary master alloys is performed, the primary phase type of the master alloys is affected by the mass ratio of Sc and Zr, and whether the morphology and distribution will have the same law. Whether the composition ratio of Sc and Zr is also one of the factors affecting the type, morphology and distribution of primary phase is still to be discussed. In addition, if the type and morphology of primary phase change, will the refinement mechanism also change?

**Author Contributions:** Conceptualization, Z.L. and S.W.; methodology, Z.L.; formal analysis, Z.L.; investigation, Z.L.; resources, S.W., H.H., W.W. and Z.N.; data curation, Z.L.; writing—original draft preparation, Z.L.; writing—review and editing, Z.L., S.W.; visualization, S.W.; supervision, S.W. and Z.N.; project administration, W.W.; funding acquisition, S.W., H.H. and Z.N. All authors have read and agreed to the published version of the manuscript.

**Funding:** National key Research and Development Program of China (Nos.2021YFB370902, 2021YFB3-704204 and 2021YFB3704205), National Natural Science Foundation of China for Innovation Research Project(No.51621003).

**Data Availability Statement:** Not applicable.

**Conflicts of Interest:** The authors declare no conflict of interest.

## References

1. McCartney, D.G. Grain refining of aluminium and its alloys using inoculants. *Int. Mater. Rev.* **1989**, *34*, 247–260. [CrossRef]
2. Ralston, K.D.; Birbilis, N. Effect of Grain Size on Corrosion: A Review. *Corrosion* **2010**, *66*, 319–324. [CrossRef]
3. Nowak, M.; Bolzoni, L.; Hari Babu, N. Grain refinement of Al–Si alloys by Nb–B inoculation. Part I: Concept development and effect on binary alloys. *Mater. Des.* **2015**, *66*, 366–375. [CrossRef]
4. Langdon, T.G. Twenty-five years of ultrafine-grained materials: Achieving exceptional properties through grain refinement. *Acta Mater.* **2013**, *61*, 7035–7059. [CrossRef]
5. Karbalaei Akbari, M.; Baharvandi, H.R.; Shirvanimoghaddam, K. Tensile and fracture behavior of nano/micro $TiB_2$ particle reinforced casting A356 aluminum alloy composites. *Mater. Des.* **2015**, *66*, 150–161. [CrossRef]
6. Wang, M.; Chen, D.; Chen, Z.; Wu, Y.; Wang, F.; Ma, N.; Wang, H. Mechanical properties of in-situ $TiB_2$/A356 composites. *Mater. Sci. Eng. A* **2014**, *590*, 246–254. [CrossRef]
7. Zhao, K.; Gao, T.; Yang, H.; Hu, K.; Liu, G.; Sun, Q.; Nie, J.; Liu, X. Enhanced grain refinement and mechanical properties of a high–strength Al–Zn–Mg–Cu–Zr alloy induced by TiC nano-particles. *Mater. Sci. Eng. A* **2021**, *806*, 140852. [CrossRef]
8. Ali, Y.; Qiu, D.; Jiang, B.; Pan, F.; Zhang, M.X. Current research progress in grain refinement of cast magnesium alloys: A review article. *J. Alloys Compd.* **2015**, *619*, 639–651. [CrossRef]
9. Wannasin, J.; Canyook, R.; Wisutmethangoon, S.; Flemings, M.C. Grain refinement behavior of an aluminum alloy by inoculation and dynamic nucleation. *Acta Mater.* **2013**, *61*, 3897–3903. [CrossRef]
10. Tamirisakandala, S.; Bhat, R.B.; Tiley, J.S.; Miracle, D.B. Grain refinement of cast titanium alloys via trace boron addition. *Scr. Mater.* **2005**, *53*, 1421–1426. [CrossRef]
11. Murty, B.S.; Kori, S.A.; Chakraborty, M. Grain refinement of aluminium and its alloys by heterogeneous nucleation and alloying. *Int. Mater. Rev.* **2002**, *47*, 3–29. [CrossRef]
12. Fan, Z.; Wang, Y.; Zhang, Y.; Qin, T.; Zhou, X.R.; Thompson, G.E.; Pennycook, T.; Hashimoto, T. Grain refining mechanism in the Al/Al–Ti–B system. *Acta Mater.* **2015**, *84*, 292–304. [CrossRef]
13. Wang, T.; Chen, Z.; Fu, H.; Xu, J.; Fu, Y.; Li, T. Grain refining potency of Al–B master alloy on pure aluminum. *Scr. Mater.* **2011**, *64*, 1121–1124. [CrossRef]
14. Zhang, M.X.; Kelly, P.M. Accurate orientation relationships between ferrite and cementite in pearlite. *Scr. Mater.* **1997**, *37*, 2009–2015. [CrossRef]
15. Zhang, M.X.; Kelly, P.M.; Easton, M.A.; Taylor, J.A. Crystallographic study of grain refinement in aluminum alloys using the edge-to-edge matching model. *Acta Mater.* **2005**, *53*, 1427–1438. [CrossRef]
16. Dai, J.; Zhu, S.; Easton, M.A.; Zhang, M.; Qiu, D.; Wu, G.; Liu, W.; Ding, W. Heat treatment, microstructure and mechanical properties of a Mg–Gd–Y alloy grain-refined by Al additions. *Mater. Sci. Eng. A* **2013**, *576*, 298–305. [CrossRef]
17. Qiu, D.; Zhang, M.X.; Taylor, J.A.; Kelly, P.M. A new approach to designing a grain refiner for Mg casting alloys and its use in Mg–Y-based alloys. *Acta Mater.* **2009**, *57*, 3052–3059. [CrossRef]
18. Luo, Y.; Zhou, L.; Kang, M.; Peng, G.; Li, W.; Miao, S.; Huang, Y.; Zhao, Y.; Xu, Y. Effect of Warm Rolling on the Grain-Refining Performance of Al-5Ti-1B Grain Refiner in Al. *JOM* **2022**, *74*, 1210–1217. [CrossRef]
19. Li, Z.; Hu, B.; Li, D.; Zhang, W.; Zeng, X.; Lin, Z.; Jin, C.; Zhao, S. Microstructure-dependent thermal conductivity and mechanical properties in cast Mg-4Sm-xAl alloys. *Mater. Sci. Eng. A* **2022**, *861*, 144336. [CrossRef]
20. Li, H.-Y.; Li, D.-W.; Zhu, Z.-X.; Chen, B.-A.; Chen, X.; Yang, C.-L.; Zhang, H.-Y.; Kang, W. Grain refinement mechanism of as-cast aluminum by hafnium. *Trans. Nonferrous Met. Soc. China* **2016**, *26*, 3059–3069. [CrossRef]
21. Chuvil'deev, V.N.; Shadrina, I.S.; Nokhrin, A.V.; Kopylov, V.I.; Bobrov, A.A.; Gryaznov, M.Y.; Shotin, S.V.; Tabachkova, N.Y.; Chegurov, M.K.; Melekhin, N.V. An investigation of thermal stability of structure and mechanical properties of Al-0.5Mg–Sc ultrafine-grained aluminum alloys. *J. Alloys Compd.* **2020**, *831*, 154805. [CrossRef]
22. Min, B.K.; Kim, H.W.; Kang, S.B. Effect of $Al_3Sc$ precipitate on the microstructural evolution during accumulative roll bonding in Al–0.2wt.% Sc alloy. *J. Mater. Process. Technol.* **2005**, *162*, 355–361. [CrossRef]
23. Knipling, K.E.; Dunand, D.C. Creep resistance of cast and aged Al–0.1Zr and Al–0.1Zr–0.1Ti (at.%) alloys at 300–400 °C. *Scr. Mater.* **2008**, *59*, 387–390. [CrossRef]

24. Knipling, K.E.; Dunand, D.C.; Seidman, D.N. Nucleation and Precipitation Strengthening in Dilute Al-Ti and Al-Zr Alloys. *Metall. Mater. Trans. A* **2007**, *38*, 2552–2563. [CrossRef]
25. Su, N.; Guan, R.; Wang, X.; Wang, Y.; Jiang, W.; Liu, H. Grain refinement in an AlEr alloy during accumulative continuous extrusion forming. *J. Alloys Compd.* **2016**, *680*, 283–290. [CrossRef]
26. Wen, S.P.; Gao, K.Y.; Li, Y.; Huang, H.; Nie, Z.R. Synergetic effect of Er and Zr on the precipitation hardening of Al–Er–Zr alloy. *Scr. Mater.* **2011**, *65*, 592–595. [CrossRef]
27. Wen, S.P.; Xing, Z.B.; Huang, H.; Huang, H.; Nie, Z.R. The effect of erbium on the microstructure and mechanical properties of Al–Mg–Mn–Zr alloy. *Mater. Sci. Eng. A* **2009**, *516*, 42–49. [CrossRef]
28. Wang, F.; Qiu, D.; Liu, Z.-L.; Taylor, J.A.; Easton, M.A.; Zhang, M.-X. The grain refinement mechanism of cast aluminium by zirconium. *Acta Mater.* **2013**, *61*, 5636–5645. [CrossRef]
29. Gao, Z.; Li, H.; Lai, Y.; Ou, Y.; Li, D. Effects of minor Zr and Er on microstructure and mechanical properties of pure aluminum. *Mater. Sci. Eng. A* **2013**, *580*, 92–98. [CrossRef]
30. Wang, W.; Pan, Q.; Lin, G.; Wang, X.; Sun, Y.; Wang, X.; Ye, J.; Sun, Y.; Yu, Y.; Jiang, F.; et al. Microstructure and properties of novel Al-Ce-Sc, Al-Ce-Y, Al-Ce-Zr and Al-Ce-Sc-Y alloy conductors processed by die casting, hot extrusion and cold drawing. *J. Mater. Sci. Technol.* **2020**, *58*, 155–170. [CrossRef]
31. Tian, S.; Li, J.; Zhang, J.; Wulabieke, Z.; Lv, D. Effect of Zr and Sc on microstructure and properties of 7136 aluminum alloy. *J. Mater. Res. Technol.* **2019**, *8*, 4130–4140. [CrossRef]
32. Jiang, Y.; Liu, F. Effects of Sc or/and Ge addition on microstructure and mechanical properties of as-cast 6016 Al alloy. *J. Alloys Compd.* **2019**, *809*, 151829. [CrossRef]
33. Ocenasek, V.; Slamova, M. Resistance to recrystallization due to Sc and Zr addition to Al–Mg alloys. *Mater. Charact.* **2001**, *47*, 157–162. [CrossRef]
34. Wang, F.; Liu, Z.L.; Qiu, D.; Taylor, J.A.; Easton, M.A.; Zhang, M.X. The Influence of the Effect of Solute on the Thermodynamic Driving Force on Grain Refinement of Al Alloys. *Metall. Mater. Trans. A Phys. Metall. Mater. Sci.* **2015**, *46*, 505–515. [CrossRef]
35. Chen, Z.w.; He, Z.; Jie, W.q. Growth restriction effects during solidification of aluminium alloys. *Trans. Nonferrous Met. Soc. China* **2009**, *19*, 410–413. [CrossRef]
36. Easton, M.A.; StJohn, D.H. A model of grain refinement incorporating alloy constitution and potency of heterogeneous nucleant particles. *Acta Mater.* **2001**, *49*, 1867–1878. [CrossRef]
37. Greer, A.L.; Bunn, A.M.; Tronche, A.; Evans, P.V.; Bristow, D.J. Modelling of inoculation of metallic melts: Application to grain refinement of aluminium by Al–Ti–B. *Acta Mater.* **2000**, *48*, 2823–2835. [CrossRef]
38. Minouei, H.; Rizi, M.S.; Akbari, G.; Hong, S.I. Effect of residual nanocrystals on thermal stability and mechanical properties of metalloid-containing amorphous alloys. *Mater. Charact.* **2021**, *173*, 110914. [CrossRef]
39. Bruce, L.B. The effect of carbide and nitride additions on the eterogeneous nucleation behavior of liquid iron. *Metall. Trans.* **1970**, *1*, 1987–1995.
40. Turnbull, D.; Vonnegut, B. Nucleation Catalysis. *Ind. Eng. Chem.* **1952**, *44*, 1292–1298. [CrossRef]
41. Zhang, M.X.; Kelly, P.M.; Ma, Q.; Taylor, J.A. Crystallography of grain refinement in Mg–Al based alloys. *Acta Mater.* **2005**, *53*, 3261–3270. [CrossRef]
42. Xiao, F.; Wu, M.; Wang, Y.; Wang, S.; Shu, D.; Wang, D.; Zhu, G.; Mi, J.; Sun, B. Design of newly effective grain refiner for aluminum based on medium-entropy metal diboride. *Vacuum* **2022**, *205*, 111462. [CrossRef]
43. Li, G.-J.; Liao, H.-C. E2EM's prediction of $LaB_6$ as nucleation substrate for primary Mn-rich phase in Al-Si-Cu-Mn heat-resistant alloy and its refining effect. *Trans. Nonferrous Met. Soc. China* **2022**, *32*, 1795–1804. [CrossRef]
44. Sunitha, K.; Gurusami, K. Study of Al-Si alloys grain refinement by inoculation. *Mater. Today Proc.* **2021**, *43*, 1825–1829. [CrossRef]
45. Greer, A.L.; Cooper, P.S.; Meredith, M.W.; Schneider, W.; Schumacher, P.; Spittle, J.A.; Tronche, A. Grain refinement of aluminium alloys by inoculation. *Adv. Eng. Mater.* **2003**, *5*, 81–91. [CrossRef]
46. Johnsson, M. Grain refinement of aluminium studied by use of a thermal analytical technique. *Thermochim. Acta* **1995**, *256*, 107–121. [CrossRef]
47. Easton, M.; StJohn, D. An analysis of the relationship between grain size, solute content, and the potency and number density of nucleant particles. *Metall. Mater. Trans. A* **2005**, *36*, 1911–1920. [CrossRef]
48. Jing, L.-J.; Pan, Y.; Lu, T.; Pi, J.-H.; Gu, T.-F. Nucleation potency prediction of LaB6 with E2EM model and its influence on microstructure and tensile properties of Al-7Si-0.3Mg alloy. *Trans. Nonferrous Met. Soc. China* **2018**, *28*, 1687–1694. [CrossRef]
49. Guan, R.G.; Misra, R.D.K.; Shang, Y.; An, Y.; Wang, Y.; Zhang, Y.; Tie, D. Mechanism of microstructural refinement of deformed aluminum under synergistic effect of $TiAl_3$ and $TiB_2$ particles and impact on mechanical properties. *Mater. Sci. Eng. A* **2018**, *716*, 129–139. [CrossRef]
50. Han, Y.; Shu, D.; Wang, J.; Sun, B. Microstructure and grain refining performance of Al-5Ti-1B master alloy prepared under high-intensity ultrasound. *Mater. Sci. Eng. A* **2006**, *430*, 326–331. [CrossRef]
51. Xu, J.; Li, R.X.; Qian, L. Effect of Agglomeration on Nucleation Potency of Inoculant Particles in the Al-Nb-B Master Alloy: Modeling and Experiments. *Metall. Mater. Trans. A* **2021**, *52*, 1077–1094. [CrossRef]
52. Quested, T.E.; Greer, A.L. The effect of the size distribution of inoculant particles on as-cast grain size in aluminium alloys. *Acta Mater.* **2004**, *52*, 3859–3868. [CrossRef]
53. Greer, A.L. Grain refinement of alloys by inoculation of melts. *Philos. Trans. R. Soc. Lond. A* **2003**, *361*, 479–495. [CrossRef]

54. Yu, H.; Wang, N.; Guan, R.; Tie, D.; Li, Z.; An, Y.; Zhang, Y. Evolution of secondary phase particles during deformation of Al-5Ti-1B master alloy and their effect on α-Al grain refinement. *J. Mater. Sci. Technol.* **2018**, *34*, 2297–2306. [CrossRef]
55. Knipling, K.E.; Dunand, D.C.; Seidman, D.N. Criteria for developing castable, creep-resistant aluminum-based alloys—A review. *Int. J. Mater. Res.* **2006**, *97*, 246–265. [CrossRef]
56. Marcantonio, J.; Mondolfo, L. Nucleation of aluminium by several intermetallic compounds. *J. Inst. Met.* **1970**, *98*, 23–27.
57. Norman, A.F.; Prangnell, P.B.; McEwen, R.S. The solidification behaviour of dilute aluminium–scandium alloys. *Acta Mater.* **1998**, *46*, 5715–5732. [CrossRef]
58. Hu, X.W.; Jiang, F.G.; Ai, F.R. Effects of rare earth Er additions on microstructure development and mechanical properties of die-cast ADC12 aluminum alloy. *J. Alloys Compd.* **2012**, *538*, 21–27. [CrossRef]
59. Muhammad, A.; Cong, X.; Wang, X.J. High strength aluminum cast alloy: A Sc modification of a standard Al–Si–Mg cast alloy. *Mater. Sci. Eng. A* **2014**, *16*, 122–126. [CrossRef]
60. Yang, J.J.; Nie, Z.R.; Jin, T.N.; Ruan, H.Q.; Zuo, T.Y. Form and refinement mechanism of element Er in Al-Zn-Mg alloy. *Chin. J. Nonferrous Met.* **2004**, *4*, 620–626. (In Chinese) [CrossRef]
61. Jones, M.J.; Humphreys, F.J. Interaction of recrystallization and precipitation: The effect of $Al_3Sc$ on the recrystallization behaviour of deformed aluminium. *Acta Mater.* **2003**, *51*, 2149–2159. [CrossRef]
62. Riddle, Y.W.; Sanders, T.H. A study of coarsening, recrystallization, and morphology of microstructure in Al-Sc-(Zr)-(Mg) alloys. *Metall. Mater. Trans. A* **2004**, *35*, 341–350. [CrossRef]
63. Yin, Z.; Pan, Q.; Zhang, Y.; Jiang, F. Effect of minor Sc and Zr on the microstructure and mechanical properties of Al–Mg based alloys. *Mater. Sci. Eng. A* **2000**, *280*, 151–155. [CrossRef]
64. Meng, Y.; ZHAO, Z.H.; Cui, J.Z. Effect of minor Zr and Sc on microstructures and mechanical properties of Al–Mg–Si–Cu–Cr–V alloys. *Trans. Nonferrous Met. Soc. China* **2013**, *23*, 1882–1889. [CrossRef]
65. Yan, K.; Chen, Z.W.; Zhao, Y.N.; Ren, C.C.; Lu, W.J.; Aldeen, A.W. Morphological characteristics of Al3Sc particles and crystallographic orientation relationships of Al3Sc/Al interface in cast Al-Sc alloy. *J. Alloys Compd.* **2021**, *861*, 158491. [CrossRef]
66. Kelly, P.; Zhang, M.-X. Edge-to-edge matching—The fundamentals. *Metall. Mater. Trans. A* **2006**, *37*, 833–839. [CrossRef]
67. Zhang, M.X.; Kelly, P.M. Edge-to-edge matching and its applications: Part II. Application to Mg–Al, Mg–Y and Mg–Mn alloys. *Acta Mater.* **2005**, *53*, 1085–1096. [CrossRef]
68. Zhang, M.X.; Kelly, P.M. Edge-to-edge matching and its applications: Part I. Application to the simple HCP/BCC system. *Acta Mater.* **2005**, *53*, 1073–1084. [CrossRef]
69. Sun, Y.; Pan, Q.; Luo, Y.; Liu, Y.; Sun, Y.; Long, L.; Li, M.; Wang, X.; Liu, S. Study on the primary $Al_3Sc$ phase and the structure heredity of Al-Zn-Mg-Cu-Sc-Zr alloy. *Mater. Charact.* **2020**, *169*, 110601. [CrossRef]
70. Hyde, K.B.; Norman, A.F.; Prangnell, P.B. The effect of cooling rate on the morphology of primary $Al_3Sc$ intermetallic particles in Al–Sc alloys. *Acta Mater.* **2001**, *49*, 1327–1337. [CrossRef]
71. Qian, M.; Stjohn, D.H. Grain nucleation and formation in Mg–Zr alloys. *Int. J. Cast Met. Res.* **2009**, *22*, 256–259. [CrossRef]
72. Qian, M.; Das, A. Grain refinement of magnesium alloys by zirconium: Formation of equiaxed grains. *Scr. Mater.* **2006**, *54*, 881–886. [CrossRef]
73. Okamoto, H. Al-Zr (aluminum-zirconium). *J. Phase Equilibria* **2002**, *23*, 455–456. [CrossRef]
74. Villars, P.; Calvert, L.D. *Pearson's Handbook of Crystallographic Data for Intermetallic Phases*; ASM: Materials Park, OH, USA, 1991.
75. Han, Y.; Ke, L.; Wang, J.; Da, S.; Sun, B. Influence of high-intensity ultrasound on grain refining performance of Al–5Ti–1B master alloy on aluminium. *Mater. Sci. Eng. A* **2005**, *405*, 306–312. [CrossRef]
76. Zhao, Y.; Zhang, S.; Chen, G.; Cheng, X. Effects of molten temperature on the morphologies of in situ $Al_3Zr$ and $ZrB_2$ particles and wear properties of $(Al_3Zr+ZrB_2)/Al$ composites. *Mater. Sci. Eng. A* **2007**, *457*, 156–161. [CrossRef]
77. Nie, Z.R.; Wen, S.P.; Huang, H.; Li, B.L.; Zuo, T.Y. Research progress of Er-containing aluminum alloy. *Chin. J. Nonferrous Met.* **2011**, *21*, 2361–2370. (In Chinese) [CrossRef]
78. Nie, Z.R.; Jin, T.N.; Zou, J.X.; Fu, J.B.; Yang, J.J.; Zuo, T.Y. Development on research of advanced rare-earth aluminum alloy. *Trans. Nonferrous Met. Soc. China* **2003**, *13*, 509–514.
79. Bai, S.; Liu, Z.; Li, Y.; Hou, Y.; Chen, X. Microstructures and fatigue fracture behavior of an Al–Cu–Mg–Ag alloy with addition of rare earth Er. *Mater. Sci. Eng. A* **2010**, *527*, 1806–1814. [CrossRef]
80. Lu, S.; Yin, D.; Zhao, Y.C.; Liu, C.; Zhao, M.C.; Yu, Z.; Wang, H.; Atrens, A. Evolution of microstructure and texture for an Al-0.4 Er alloy during accumulative roll bonding. *J. Alloys Compd.* **2019**, *811*, 152005. [CrossRef]
81. Yang, J.J.; Nie, Z.R.; Jin, T.N.; Xu, G.F.; Fu, J.B.; Ruan, H.Q.; Zuo, T.Y. Effect of trace rare earth element Er on high pure Al. *Trans. Nonferrous Met. Soc. China* **2003**, *5*, 1035–1039.
82. Mondolfo, L.F. Al–Er Aluminum–Erbium system. In *Aluminum Alloys*; Butterworth-Heinemann: Oxford, UK, 1976; p. 280. [CrossRef]
83. Zou, L.; Pan, Q.-L.; He, Y.-B.; Wang, C.-Z.; Liang, W.-J. Effect of minor Sc and Zr addition on microstructures and mechanical properties of Al-Zn-Mg-Cu alloys. *Trans. Nonferrous Met. Soc. China* **2007**, *17*, 340–345. [CrossRef]
84. Esmaeili Ghayoumabadi, M.; Mochugovskiy, A.G.; Tabachkova, N.Y.; Mikhaylovskaya, A.V. The influence of minor additions of Y, Sc, and Zr on the microstructural evolution, superplastic behavior, and mechanical properties of AA6013 alloy. *J. Alloys Compd.* **2022**, *900*, 163477. [CrossRef]

85. Knipling, K.E.; Seidman, D.N.; Dunand, D.C. Ambient- and high-temperature mechanical properties of isochronally aged Al–0.06Sc, Al–0.06Zr and Al–0.06Sc–0.06Zr (at.%) alloys. *Acta Mater.* **2011**, *59*, 943–954. [CrossRef]
86. Lee, S.; Utsunomiya, A.; Akamatsu, H.; Neishi, K.; Furukawa, M.; Horita, Z.; Langdon, T.G. Influence of scandium and zirconium on grain stability and superplastic ductilities in ultrafine-grained Al–Mg alloys. *Acta Mater.* **2002**, *50*, 553–564. [CrossRef]
87. Yuzbekova, D.; Mogucheva, A.; Kaibyshev, R. Superplasticity of ultrafine-grained Al–Mg–Sc–Zr alloy. *Mater. Sci. Eng. A* **2016**, *675*, 228–242. [CrossRef]
88. Wang, Y.; Zhang, Z.; Wu, R.; Sun, J.; Jiao, Y.; Hou, L.; Zhang, J.; Li, X.; Zhang, M. Ambient-temperature mechanical properties of isochronally aged 1420-Sc-Zr aluminum alloy. *Mater. Sci. Eng. A* **2019**, *745*, 411–419. [CrossRef]
89. Kuang, Q.; Wang, R.; Peng, C.; Cai, Z. Comparison of microstructure and mechanical properties of Al-Mg-Li-Sc-Zr alloys processed by ingot metallurgy and rapid solidification. *J. Alloys Compd.* **2021**, *883*, 160937. [CrossRef]
90. Zhang, X.; Zhang, L.; Wu, G.; Sun, J.; Rong, M.; Hsieh, C.-C.; Yu, Y. Influence of Sc content on the microstructure and mechanical properties of cast Al-2Li-2Cu-0.5Mg-0.2Zr alloy. *Mater. Charact.* **2018**, *142*, 223–236. [CrossRef]
91. Fuller, C.B.; Seidman, D.N.; Dunand, D.C. Mechanical properties of Al(Sc,Zr) alloys at ambient and elevated temperatures. *Acta Mater.* **2003**, *51*, 4803–4814. [CrossRef]
92. Zhao, C.; Li, Y.; Xu, J.; Luo, Q.; Jiang, Y.; Xiao, Q.; Li, Q. Enhanced grain refinement of Al-Si alloys by novel Al-V-B refiners. *J. Mater. Sci. Technol.* **2021**, *94*, 104–112. [CrossRef]

MDPI AG
Grosspeteranlage 5
4052 Basel
Switzerland
Tel.: +41 61 683 77 34

*Metals* Editorial Office
E-mail: metals@mdpi.com
www.mdpi.com/journal/metals